PELICAN BOOKS

THE ELIZABETHAN WORLD PICTURE

E. M. W. Tillyard, Litt.D., who was Master of Jesus College, Cambridge, from 1945 to 1959, and a university lecturer in English for nearly thirty years, ranked among the leading authorities on both Shakespeare and Milton. He was born in 1889 and attended a school in Lausanne and the Perse School, Cambridge, before proceeding to Jesus College, where he gained a Double First in Classics and was Craven Student. After studying archaeology in Athens, he returned as a fellow to Jesus College in 1913. He was on active service in France and at Salonika during the First World War, at the end of which he was a liaison officer with the Greek forces. After the Second World War he was a special lecturer at three transatlantic universities. His publications include *Milton*, *Shakespeare's Last Plays*, *Shakespeare's History Plays*, *The Miltonic Setting*, *The Elizabethan World Picture*, *Shakespeare's Problem Plays*, *The Metaphysicals and Milton*, and *The Epic Strain in the English Novel*. Dr Tillyard, who was married and had three children, died in 1962.

THE
ELIZABETHAN
WORLD
PICTURE

—

E. M. W. TILLYARD

—

PENGUIN BOOKS
in association with Chatto & Windus

Penguin Books Ltd, Harmondsworth, Middlesex, England
Penguin Books Australia Ltd, Ringwood, Victoria, Australia

—

First published by Chatto & Windus 1943
Published in Peregrine Books 1963
Reprinted 1966, 1968, 1970
Reissued in Pelican Books 1972

—

Copyright © Stephen Tillyard, 1943

—

Made and printed in Great Britain
by Cox & Wyman Ltd,
London, Reading and Fakenham
Set in Monotype Bembo

CONTENTS

PREFACE

THIS small book has come out of an attempt to write a larger one on Shakespeare's Histories. In studying these I concluded that the picture of civil war and disorder they present had no meaning apart from a background of order to judge them by. My first chapter set out to describe that background. When it was finished, I found that it applied to Shakespeare's Histories no more than to the rest of Shakespeare or indeed than to Elizabethan literature generally. I also found that the order I was describing was much more than political order, or, if political, was always a part of a larger cosmic order. I found, further, that the Elizabethans saw this single order under three aspects: a chain, a set of correspondences, and a dance. Here then was a subject too big for a single chapter in a more specialized book, a subject demanding separate treatment.

Now this idea of cosmic order was one of the genuine ruling ideas of the age, and perhaps the most characteristic. Such ideas, like our everyday manners, are the least disputed and the least paraded in the creative literature of the time. The Victorians believed in the virtue of self-help, yet we do not associate the poems of Tennyson or the novels of George Eliot with the belief. They take it too much for granted. Of course if we read these works with the idea in our minds we shall find abundant hints of it. And to be ignorant of it makes us less able to understand these two authors. The province of this book is some of the notions about the world and man which were quite taken for granted by the ordinary educated Elizabethan; the utter commonplaces too familiar for the poets to make detailed use of except in explicitly didactic passages, but essential as basic assumptions and invaluable at moments of high passion. Shakespeare glances at one of these essential commonplaces when, in *Julius Caesar*, he makes Brutus compare the state of man to a little kingdom. The comparison of man to the state or 'body politic' was as fundamental to the Elizabethans as the belief in self-help was to the Victorians.

My object then is to extract and expound the most ordinary beliefs about the constitution of the world as pictured in the Elizabethan age and through this exposition to help the ordinary reader to understand and to enjoy the great writers of the age. In attempting this I have incidentally brought together a number of pieces of elementary lore which I have not found assembled elsewhere. This book may actually be a convenient factual aid to the bare construing of some of Spenser or Donne or Milton.

Though I have mainly expounded, I have sometimes drawn conclusions, and I have illustrated the way a belief passed into the literature of the time. As I write for the ordinary reader not the specialist I have used the best-known writers for such illustration. On the other hand when I am setting forth an actual doctrine I do not avoid illustrating from unfamiliar writers. It has been impossible always to distinguish between these two kinds of illustration; and the reader must not be surprised if he finds a piece of Shakespeare or Milton used simultaneously to state a doctrine and to exemplify the use poetry can make of it.

I must warn readers that some of the facts are only approximate. There were many variations of opinion about the way the universe was constituted impossible to record in a short book. I have done my best to choose always the most usual opinion. If any specialist in the period reads this book, I hope he may agree with me that the doctrines I have expounded are all sufficiently commonplace and may find that as few as possible of the relevant commonplaces have escaped me.

It is unfortunate that the facts with which I have to deal, though all equally familiar to an Elizabethan, are not so to a modern. A part, like the four humours, is familiar, even to distress; but a part, like the notion of the 'vast chain of being', will be new to the ordinary reader. As in a short account proportion is everything, I cannot allow degrees of familiarity to dictate the space or the emphasis I give to different matters. First things must have first place. And if I speak of stale things as if they were fresh and obscure things as if they were known, it is to preserve the proportions in which I imagine the Elizabethans saw them all.

In quoting I have thought of the ordinary reader's convenience and have modernized spelling and punctuation, except for Milton. Milton took great care over these things and hardly suffers in intelligibility from having them preserved.

I sometimes use the word Elizabethan with great laxity, meaning anything within the compass of the English Renaissance, anything between the ages of Henry VIII and Charles I akin to the main trends of Elizabethan thought.

My thanks are due to friends who have put me on to references I might have missed: to Miss E. E. H. Welsford, M.A., Fellow of Newnham College, to Miss R. Freeman, PH.D., Girton College, Lecturer at Birkbeck College, to Professor Theodore Spencer, PH.D., of Trinity College Cambridge and Harvard University, and to Mr Donald Gordon, PH.D., of Edinburgh University and Trinity College Cambridge.

Finally I must pay my tribute to recent American work on Renaissance thought; work the cumulative magnitude of which is not always recognized in England. I mean, for instance, that of the late Edwin Greenlaw and his associates or of Professor Charles G. Osgood and the other editors of the Variorum Spenser. Without this work I should not have dared to generalize as I have done.

I regret that Professor Theodore Spencer's *Shakespeare and the Nature of Man* (New York, 1942) reached me after my text was in type. We have been writing, independently, of some of the same things, and I wish I could have made many references to this book. All I can do now is refer generally to the learning and the charm with which he develops his theme.

Jesus College   E.M.W.T.
Cambridge

INTRODUCTORY

PEOPLE still think of the Age of Elizabeth as a secular period between two outbreaks of Protestantism: a period in which religious enthusiasm was sufficiently dormant to allow the new humanism to shape our literature. They admit indeed that the quiet was precarious and that the Puritans were ever on the alert. But they allow the emphasis to be on the Queen's political intuitions, the voyages of discovery, and the brilliant externals of Elizabethan life. The first pages of Virginia Woolf's *Orlando* are in these matters typical. They do not tell us that Queen Elizabeth translated Boethius, that Raleigh was a theologian as well as a discoverer, and that sermons were as much a part of an ordinary Elizabethan's life as bear-baiting. The way Hamlet's words on man are often taken will illustrate this habit of mind.

What a piece of work is a man: how noble in reason; how infinite in faculty; in form and moving how express and admirable; in action how like an angel; in apprehension how like a god; the beauty of the world, the paragon of animals.

This has been taken as one of the great English versions of Renaissance humanism, an assertion of the dignity of man against the asceticisms of medieval misanthropy. Actually it is in the purest medieval tradition: Shakespeare's version of the orthodox encomia of what man, created in God's image, was like in his prelapsarian state and of what ideally he is still capable of being. It also shows Shakespeare placing man in the traditional cosmic setting between the angels and the beasts. It was what the theologians had been saying for centuries. Here is a typical version, by Nemesius, a Syrian bishop of the fourth century, in George Wither's translation:

No eloquence may worthily publish forth the manifold pre-eminences and advantages which are bestowed on this creature. He passeth over the vast seas; he rangeth about the wide heavens by his contemplation and conceives the motions and magnitudes of the stars . . . He is learned in every science and skilful in artificial workings . . . He talketh with angels yea with God himself. He hath all the creatures within his dominion.

What is true of Hamlet on man is in the main true of Elizabethan modes of thought in general.

The thing that *Orlando* (and for that matter *Shakespeare's England* taken all in all) misses is that the Puritans and the courtiers were more united by a common theological bond than they were divided by ethical disagreements. They had in common a mass of basic assumptions about the world, which they never disputed and whose importance varied inversely with this very meagreness of controversy.

Coming to the world picture itself, one can say dogmatically that it was still solidly theocentric, and that it was a simplified version of a much more complicated medieval picture. Now the Middle Ages derived their world picture from an amalgam of Plato and the Old Testament, invented by the Jews of Alexandria and vivified by the new religion of Christ. It was unlike paganism (apart from Platonism and some mystery cults) in being theocentric, and it resembled Platonism and other theocentric cults in being perpetually subjected to the conflicting claims of this and another world. To hold that the other world, because so persistently advertised, had it all its own way in the experience of medieval thinkers is as simple-minded as to hold that all Germans are merciless because their leaders have ordered them to be so, or that England must have indeed been merry between the two wars because of all the incitements by theatre or wayside pulpit to be cheerful. On reflection we can only conclude that many Germans must be obstinately kind to need such orders and that many Englishmen refused to be comforted to need such advice. Those who know most about the Middle Ages now assure us that humanism and a belief in the present life were powerful by the twelfth century, and that exhortations to contemn the world were themselves powerful at that time for that very reason. The two contradictory principles coexisted in a state of high tension. Further it is an error to think that with the Renaissance the belief in the present life won a definitive victory. Till recently Petrarch's imaginary dialogue between himself and St Augustine, known as his *Secret*, was thought to typify the transition from Middle Ages to Renaissance because it deals with this same conflict as if there might be a doubt about the result. Actually it does not differ greatly in spirit from the most popular of all moral treatises during the Middle Ages, the dialogue Boethius held between himself and Divine Philosophy; it shows no slackening of ardour in presenting the old arguments for despising the world. Indeed from Augustine himself through the Middle Ages and the Renaissance, through the Elizabethans to Donne and Milton, the old arguments persisted. The Duke's exhortation to Claudio in *Measure for Measure*, 'Be absolute for death', is an epitome of medieval homilies on the contempt of the world. And when Boethius calls the love of fame 'the one thing that could allure geniuses outstanding but not yet quite perfected in virtue' and Milton calls it 'that last infirmity of noble mind', the truth is not that Milton was copying Boethius but that he was giving his own version of the perpetual struggle. The conclusion then is that, though there were various new things in the Elizabethan age to make life exciting, the old struggle between the claims of two worlds persisted and that to look on this age as mainly secular is wrong.

The world picture which the Middle Ages inherited was that of an ordered universe arranged in a fixed system of hierarchies but modified by man's sin and the hope of his redemption. The same energy that carried through their feats of architecture impelled them to elaborate this inherited picture. Everything had to be included and everything had to be made to fit and to connect. For instance, it would not do to enjoy the *Aeneid* as the epic of Augustan Rome: the poem had to be fitted into the current theological scheme and was interpreted as an allegory of the human soul from birth to death. Once invented, the conventions of courtly love had to be given their precise value in the total scheme. Thus Launcelot, the perfect courtly lover, is the champion of chivalry but is denied the vision of the Grail: the limits of his possible virtue are precisely set.

Typical of much medieval elaboration and precise correspondence of detail was the habit of acting in accordance with the position of the planets. There is a good popular exposition of this in the first chapter of J. L. Lowes's *Geoffrey Chaucer*. From the astronomer Ptolemy the Middle Ages derived the custom of associating certain classes of people with certain planets. Further, they allotted to a single planet every hour of the week. What use they put these notions to can be seen in the third part of Chaucer's *Knight's Tale*, where Palamon Emily and Arcite visit the temples of Venus Diana and Mars. This they do at exactly the hours appropriated to these planetary divinities. For Palamon to have prayed to Venus in an hour appropriated to Mercury would be profane indecent and perilous. In the fourth book of Ptolemy's astronomical treatise we read that Mars, in aspect with the sun, makes his subjects those who use fire in their crafts, such as cooks moulders cauterizers smiths workers in mines. Hence, when on the temple of Mars Chaucer puts the picture of

> The cook y-scalded for al his longe ladel,

he is being scrupulously correct.

One is tempted to call the medieval habit of life mathematical or to compare it with a gigantic game where everything is included and every act is conducted under the most complicated system of rules.

Ultimately the game grew over-complicated and was too much for people. But it is a mistake to think that it was changed. Protestantism was largely a selection and a simplification of what was there all the time. It mattered little to the sublime scheme of Augustinian and Thomist theology if indulgences for instance were done away with. And the universe was still an order, even if men forgot many of the details of its internal concatenations. The kind of thing that had happened can be seen by comparing the above Chaucerian scene with its parallel in the *Two Noble Kinsmen*, almost certainly an example of Shakespeare's very latest work. Here there is no trace of Chaucer's astronomical detail. Yet it mentions, and with effective solemnity, one of the great medieval commonplaces, the one already hinted at in Brutus's speech. Arcite prays to Mars:

> O great corrector of enormous times,
> Shaker of o'er-rank states, thou grand decider
> Of dusty and old titles, that heal'st with blood
> The earth when it is sick, and cur'st the world
> O' the plurisy of people.

Here war is presented as part of the great cosmic scene and is secured in its place by being to the body politic as the medical operation of blood-letting is to the human body. Not that many details of the old correspondences did not linger on: they did but often, like the Punch and Judy show today, bereft of ancestral dignity. This, for instance, from *Twelfth Night* is what survives in Shakespeare of the medieval correspondence between parts of the body and the constellations:

SIR TOBY BELCH: I did think by the excellent constitution of thy leg it was formed under the star of a galliard.

SIR ANDREW AGUECHEEK: Ay, 'tis strong, and it does indifferently well in a flame-coloured stock. Shall we set about some revels?

SIR TOBY: What shall we do else? were we not born under Taurus?

SIR ANDREW: Taurus: that's sides and heart.

SIR TOBY: No, sir, it is legs and thighs.

Characteristically both speakers are made to get the association wrong; and Shakespeare probably knew that to Taurus were assigned the neck and throat. There is irony in Sir Toby's being right in a way he did not mean. He meant to refer to dancing – legs and thighs – but the drinking implied by neck and throat is just as apt to the proposed revels. The present point is that the serious and ceremonious game of the Middle Ages has degenerated into farce.

But though the general medieval picture of the world survived in outline into the Elizabethan age, its existence was by then precarious. There had been Machiavelli, to whom the idea of a universe divinely ordered throughout was repugnant, and in the seventeenth century men began to understand and heed and not merely to travesty and abuse him. Recent research has shown that the educated Elizabethan had plenty of textbooks in the vernacular instructing him in the Copernican astronomy, yet he was loath to upset the old order by applying his knowledge. The new commercialism was hostile to medieval stability. The greatness of the Elizabethan age was that it contained so much of new without bursting the noble form of the old order. It is here that the Queen herself comes in. Somehow the Tudors had inserted themselves into the constitution of the medieval universe. They were part of the pattern and they made themselves indispensable. If they were to be preserved, it had to be as part of this pattern. It was a serious matter not a mere fancy if an Elizabethan writer compared Elizabeth to the *primum mobile*, the master-sphere of the physical universe, and every activity within the realm to the varied motions of the other spheres governed to the last fraction by the influence of their container.

2

ORDER

THOSE (and they are at present the majority) who take their notion of the Elizabethan age principally from the drama will find it difficult to agree that its world picture was ruled by a general conception of order, for at first sight that drama is anything but orderly. However, people are beginning to perceive that this drama was highly stylized and conventional, that its technical licences are of certain kinds and fall into a pattern, that its extravagant sentiments are repetitions and not novelties; that it may after all have its own, if queer, regulation. Actually the case is such as I have described in my preface: the conception of order is so taken for granted, so much part of the collective mind of the people, that it is hardly mentioned except in explicitly didactic passages. It is not absent from non-didactic writing, for it appears in Spenser's *Hymns of Love* and in Ulysses' speech on 'degree' in Shakespeare's *Troilus and Cressida*. It occurs frequently in didactic prose: in Elyot's *Governor*, the Church Homily *Of Obedience*, the first book of Hooker's *Laws of Ecclesiastical Polity*, and the preface to Raleigh's *History of the World*. Shakespeare's version is the best known. For this reason, and because its full scope is not always perceived, I begin with it.

> The heavens themselves, the planets, and this centre
> Observe degree priority and place
> Insisture course proportion season form
> Office and custom, in all line of order;
> And therefore is the glorious planet Sol
> In noble eminence enthron'd and spher'd
> Amidst the other, whose med'cinable eye
> Corrects the ill aspects of planets evil
> And posts like the commandment of a king,

Sans check, to good and bad. But when the planets
In evil mixture to disorder wander,
What plagues and what portents, what mutiny,
What raging of the sea, shaking of earth,
Commotion in the winds, frights changes horrors,
Divert and crack, rend and deracinate
The unity and married calm of states
Quite from their fixture. Oh, when degree is shak'd,
Which is the ladder to all high designs,
The enterprise is sick. How could communities,
Degrees in schools and brotherhoods in cities,
Peaceful commerce from dividable shores,
The primogenitive and due of birth,
Prerogative of age, crowns sceptres laurels,
But by degree stand in authentic place?
Take but degree away, untune that string,
And hark, what discord follows. Each thing meets
In mere oppugnancy. The bounded waters
Should lift their bosoms higher than the shores
And make a sop of all this solid globe.
Strength should be lord to imbecility,
And the rude son should strike his father dead.
This chaos, when degree is suffocate,
Follows the choking.

Much of what I have to expound is contained in this passage,
and I shall revert to its details later. The point here is that so
many things are included simultaneously within this 'degree'
or order, and so strong a sense is given of their interconnections.
The passage is at once cosmic and domestic. The sun, the king,
primogeniture hang together; the war of the planets is echoed
by the war of the elements and by civil war on earth; the
homely brotherhoods or guilds in cities are found along with
an oblique reference to creation out of the confusion of chaos.
Here is a picture of immense and varied activity, constantly
threatened with dissolution, and yet preserved from it by a
superior unifying power. The picture, however, though so
rich, is not complete. There is nothing about God and the

angels, nothing about animals vegetables and minerals. For Shakespeare's dramatic purposes he brought in quite enough, but it would be wrong to think that he did not mean to imply the two extremes of creation also or that he would have disclaimed the following account of 'degree': Raleigh, after enlarging on the joys of heaven, which will make any earthly joy negligible, continues,

Shall we therefore value honour and riches at nothing and neglect them as unnecessary and vain? Certainly no. For that infinite wisdom of God, which hath distinguished his angels by degrees, which hath given greater and less light and beauty to heavenly bodies, which hath made differences between beasts and birds, created the eagle and the fly, the cedar and the shrub, and among stones given the fairest tincture to the ruby and the quickest light to the diamond, hath also ordained kings, dukes or leaders of the people, magistrates, judges, and other degrees among men.

One of the clearest expositions of order (and close to Shakespeare's though a good deal earlier in date) is Elyot's in the first chapter of the *Governor*. It has this prominent place because order is the condition of all that follows; for of what use to educate the magistrate without the assurance of a coherent universe in which he can do his proper work?

Take away order from all things, what should then remain? Certes nothing finally, except some man would imagine eftsoons chaos. Also where there is any lack of order needs must be perpetual conflict. And in things subject to nature nothing of himself only may be nourished; but, when he hath destroyed that wherewith he doth participate by the order of his creation, he himself of necessity must then perish; whereof ensueth universal dissolution.

Hath not God set degrees and estates in all his glorious works? First in his heavenly ministers, whom he hath constituted in divers degrees called hierarchies. Behold the four elements, whereof the body of man is compact, how they be set in their places called spheres, higher or lower according to the sovereignty of their natures. Behold also the order that God hath put generally in all his creatures, beginning at the most inferior or base and ascending upward. He made not only herbs to garnish the earth but also trees of a more eminent stature than herbs.

Semblably in birds beasts and fishes some be good for the sustenance
of man, some bear things profitable to sundry uses, other be apt to
occupation and labour. Every kind of trees herbs birds beasts and
fishes have a peculiar disposition appropered unto them by God their
creator; so that in everything is order, and without order may be
nothing stable or permanent. And it may not be called order except it
do contain in it degrees, high and base, according to the merit or
estimation of the thing that is ordered.

This is all very explicit and prosaic. It is what everyone believed
in Elizabeth's days and it is *all* there behind such poetic statements of order as the following from Spenser's *Hymn of Love*
describing creation:

> The earth the air the water and the fire
> Then gan to range themselves in huge array
> And with contrary forces to conspire
> Each against other by all means they may,
> Threat'ning their own confusion and decay:
> Air hated earth and water hated fire,
> Till Love relented their rebellious ire.
>
> He then them took and, tempering goodly well
> Their contrary dislikes with loved means,
> Did place them all in order and compel
> To keep themselves within their sundry reigns
> Together linkt with adamantine chains:
> Yet so as that in every living wight
> They mix themselves and show their kindly might.
>
> So ever since they firmly have remained
> And duly well observed his behest,
> Through which now all these things that are contained
> Within this goodly cope, both most and least,
> Their being have.

The conception of order described above must have been
common to all Elizabethans of even modest intelligence.
Hooker's elaborated account must have stated pretty fairly the

preponderating conception among the educated. Hooker is not easy reading to a modern but would have been much less difficult to a contemporary used to his kind of prose. He writes not for the technical theologian but mediates theology to the general educated public of his day. He is master of the sort of summary which, though it avoids irksome and controversial detail, presents the general and the simplified with consummate force and freshness. He has the acutest sense of what the ordinary educated man can grasp and having grasped ratify. It is this tact that assures us that he speaks for the educated nucleus that dictated the current beliefs of the Elizabethan Age. He represents far more truly the background of Elizabethan literature than do the coney-catching pamphlets or the novel of low life.

Hooker's version is of course avowedly theological and it is more explicit, but the order it describes is Elyot's and Shakespeare's. His name for it is law, law in its general sense. Above all cosmic or earthly orders or laws there is Law in general, 'that Law which giveth life unto all the rest which are commendable just and good, namely the Law whereby the Eternal himself doth work'. By a masterly ambiguity he avoids the great traditional dispute whether a thing is right because God wills it, or God wills it because it is right. God created his own law both because he willed it and because it was right. Though voluntary it was not arbitrary, but based on reason. That divine reason is beyond our understanding; yet we know it is there. God's law is eternal, 'being that order which God before all ages hath set down with himself, for himself to do all things by'. God chose to work in finitude in some sort to show his glory; and having so chosen he expressed the abundance of his glory in variety. The sense of full life given by Shakespeare's 'degree' speech is a close poetical parallel to this theological doctrine of variety. From the single generating law of God Hooker goes on to describe the subordinate and separate laws; for law too must become multiple when it is applied to an abundantly diversified creation. God, as well as creating his own eternal law, issued his command in accordance with it:

That part of it which ordereth natural agents we call usually nature's law; that which angels do clearly behold and without any swerving observe is a law celestial and heavenly; the law of reason, that which bindeth creatures reasonable in this world and with which by reason they may most plainly perceive themselves bound; that which bindeth them and is not known but by special revelation from God, divine law; human law, that which out of the law either of reason or of God men probably gathering to be expedient, they make it a law.

Hooker's first book comes to rest in a final summary, which includes the notion of law or order as harmony ('Take but degree away, untune that string, And hark what discord follows.'):

Wherefore that here we may briefly end: of law there can be no less acknowledged than that her seat is the bosom of God, her voice the harmony of the world: all things in heaven and earth do her homage, the very least as feeling her care and the greatest not exempted from her power; both angels and men and creatures of what condition soever, though each in differing sort and manner yet all with uniform consent, admiring her as the mother of their peace and joy.

Though little enlarged on by the poets, cosmic order was yet one of the master-themes of Elizabethan poetry. It has its positive and its negative expressions. First there is an occasional full statement, as in Spenser's *Hymns*. Then there are the partial statements or the hints. Ulysses' 'degree' speech is a partial statement. The long scene between Malcolm and Macduff at the English court and the reference to the healing power of the English king draw their strength from the idea. There is a short passage in the first part of *Henry VI* whose pivotal meaning any other than a contemporary reader might easily miss. It shows Talbot during a truce with the French doing homage to Henry VI, who has arrived at Paris to be crowned, and Henry rewarding him with the earldom of Shrewsbury. The scene is an example of the sort of thing that ought to happen in an orderly kingdom and it serves as a norm by which the many disorders in the same play are judged. Talbot's speech in its references to the places of God, the king, and himself in their due degrees carries with it the whole context of Hooker and the great Homily of obedience:

> My gracious prince and honorable peers,
> Hearing of your arrival in this realm,
> I have awhile given truce unto my wars,
> To do my duty to my sovereign.
> In sign whereof this arm, that hath reclaim'd
> To your obedience fifty fortresses
> Twelve cities and seven walled towns of strength,
> Beside five hundred prisoners of esteem,
> Lets fall his sword before your highness' feet
> And with submissive loyalty of heart
> Ascribes the glory of his conquest got
> First to my God and next unto your grace.

The gorgeous emblematical figure of Ceremony coming to rebuke the lawless loves of Hero and Leander in Chapman's continuation of Marlowe's poem is yet another, and far more explicit and academic, version:

> The goddess Ceremony, with a crown
> Of all the stars . . .
> Her flaming hair to her bright feet descended,
> By which hung all the bench of deities.
> And in a chain, compact of ears and eyes,
> She led Religion. All her body was
> Clear and transparent as the purest glass,
> For she was all presented to the sense.
> Devotion Order State and Reverence
> Her shadows were.

The notion of cosmic order pervades the entire *Fairy Queen* and prompts such a detail as Spenser's iteration of the phrase 'in comely rew [row]' or 'on a row'. The arrangement is comely not just because it is pretty and seemly but because it harmonizes with a universal order.

But the negative implication was even more frequent and emphatic. If the Elizabethans believed in an ideal order animating earthly order, they were terrified lest it should be upset, and appalled by the visible tokens of disorder that suggested its upsetting. They were obsessed by the fear of chaos and the fact of mutability; and the obsession was powerful in proportion as their faith in the cosmic order was strong. To us *chaos* means hardly more than confusion on a large scale; to an Elizabethan it meant the cosmic anarchy before creation and the wholesale dissolution that would result if the pressure of Providence relaxed and allowed the law of nature to cease functioning. Othello's 'chaos is come again', or Ulysses' 'this chaos, when degree is suffocate', cannot be fully felt apart from orthodox theology. Hooker's own description of chaos gives the proper context:

Now if nature should intermit her course and leave altogether, though it were but for a while, the observation of her own laws; if those principal and mother elements of the world, whereof all things in this lower world are made, should lose the qualities which now they have; if the frame of that heavenly arch erected over our heads should loosen and dissolve itself; if celestial spheres should forget their wonted motions and by irregular volubility turn themselves any way as it might happen; if the prince of the lights of heaven, which now as a giant doth run his unwearied course, should as it were through a languishing faintness begin to stand and to rest himself; if the moon should wander from her beaten way, the times and seasons of the year blend themselves by disordered and confused mixture, the winds breathe out their last gasp, the clouds yield no rain, the earth be defeated of heavenly influence, the fruits of the earth pine away as children at the withered breasts of their mother no longer able to yield them relief: what would become of man himself, whom these things now do all serve? See we not plainly that obedience of creatures unto the law of nature is the stay of the whole world?

If Shakespeare in *Henry VI*, *Troilus and Cressida*, and *Macbeth* gives us his version of order, it bulks small compared with the different kinds of chaos that reign or threaten in all these plays. Yet Shakespeare's chaos is without meaning apart from the proper background of cosmic order by which to judge it.

While Shakespeare puts the opposition to order and his desire for it in terms of chaos mainly, Spenser (and above all in the *Fairy Queen*) does so in terms of mutability. In the Garden of Adonis, the generative workshop of nature, time, which works changes, is the great enemy; and in the last stanzas of the poem the goddess Mutability claims sway in the world. Like Boethius at the beginning of his second book of the *Consolation of Philosophy* Spenser concludes that Mutability is part of a larger stability, just as the wind shows its constancy in never failing to be changeable; but it is through the poignancy of his regrets for earthly instability that he so wonderfully expresses his overmastering passion for order and the old double pull to relish the world and to despise it:

When I bethink me on that speech whilere
Of Mutability and well it weigh,
Me seems that though she all unworthy were
Of the heav'n's rule, yet, very sooth to say,
In all things else she bears the greatest sway:
Which makes me loathe this state of life so tickle
And love of things so vain to cast away;
Whose flow'ring pride, so fading and so fickle,
Short time shall soon cut down with his consuming sickle.

Then gin I think on that which Nature said
Of that same time when no more change shall be.
But stedfast rest of all things, firmly stay'd
Upon the pillars of eternity,
That is contrare to mutability:
For all that moveth doth in change delight:
But henceforth all shall rest eternally
With him that is the God of Sabaoth hight.
O, that great Sabaoth God, grant me that Sabbath's sight.

3

SIN

THE conception of world order was for the Elizabethans a principal matter; the other set of ideas that ranked with it was the theological scheme of sin and salvation. However widely biblical history was presented in medieval drama and sermon, even though many Protestants may have had the Gospels by heart, the part of Christianity that was paramount was not the life of Christ but the orthodox scheme of the revolt of the bad angels, the creation, the temptation and fall of man, the incarnation, the atonement, the regeneration through Christ. And this is as true of the Middle Ages as of the age of Elizabeth. Here was another pattern as powerful in its imaginative appeal as the divine order of the universe, standing out clean and distinct from all other theological lore. It had been the pattern St Paul imposed on the Christian material and it persisted in spite of all disputes and eruptions as a great formative force in men's lives, until with the scientific revolution of the seventeenth century it slowly receded into this and that citadel. We should never let ourselves forget that the orthodox scheme of salvation was pervasive in the Elizabethan age. You could revolt against it but you could not ignore it. Atheism not agnosticism was the rule. It was far easier to be very wicked and think yourself so than to be a little wicked without a sense of sin.

There are so few references to the Pauline scheme of redemption in the sonneteers and dramatists that this insistence on its being essential to the Elizabethan world picture might well be disputed. Yet this very scarcity is a sign of extreme familiarity, and even a single reference will be vast in its implications. Take these words of Angelo and Isabella arguing about Claudio's condemnation in *Measure for Measure*:

ANGELO:
>   Your brother is a forfeit of the law,
>   And you but waste your words.

ISABELLA: Alas, alas!
>   Why, all the souls that were were forfeit once;
>   And He that might the vantage best have took
>   Found out the remedy.

The reference is of the slightest, yet it reveals and takes for
granted the total Pauline theology of Christ abrogating the
enslavement to the old law incurred through the defection of
Adam. Outside the sonneteers and the dramatists the case can
be very different. In Spenser's *Hymns* (which as a whole are a
pretty fair epitome of what this and the previous chapter are
concerned with) the scheme of salvation is given a principal
place:

> Out of the bosom of eternal bliss,
> In which he reigned with his glorious sire,
> He down descended, like a most demiss
> And abject thrall, in flesh's frail attire,
> That he for him might pay sin's deadly hire
> And him restore unto that happy state,
> In which he stood before his hapless fate.
>
> In flesh at first the guilt committed was;
> Therefore in flesh it must be satisfied:
> Nor spirit nor angel, though they man surpass,
> Could make amends to God for man's misguide,
> But only man himself, who self did slide.
> So, taking flesh of sacred virgin's womb,
> For man's dear sake he did a man become.

When it comes to indirect manifestation it is difficult to
speak. English literature as a whole has spoken an idiom
permeated by Christian dogma; in specifying a passage in
Elizabethan literature showing that idiom one may be pointing
to something the very reverse of remarkable. But indeed all I
am trying to establish is that the Elizabethan age is of a piece
with what went before and after it, that it is *not* remarkable for

its departure from the norm. Out of so many possible examples take the *Duchess of Malfi*. We know nothing of Webster's beliefs, but it matters little what he professed in comparison with what he betrays. The Duchess herself, wilfully courting her steward against all propriety and 'degree', knows she is erring, that she is going into a wilderness. Webster may personally wish to justify her but he cannot free her from the context of sin and its atonement. Bosola too is in the same context. He belongs to a world of violent crime and violent change, of sin blood and repentance, yet to a world loyal to a theological scheme. Indeed all the violence of Elizabethan drama has nothing to do with a dissolution of moral standards: on the contrary, it can afford to indulge itself just because those standards were so powerful. Men were bitter and thought the world was in decay largely because they expected so much. It is the sententious protests of Restoration drama that first betray a moral weakening.

I have written of order and of sin and redemption separately, but in practice the two schemes were fused, just as in Milton's poem *On Time* disproportionated sin jars against nature's chime, within the limits of a grammatical clause. There is the everpresent glory of God's creation, the perpetual pressure of his Providence. Yet disorder or chaos, the product of sin, is perpetually striving to come again. And if, by tradition, the way to salvation is through God's grace and Christ's atonement, there is also the way, paired with it, through the contemplation of the divine order of the created universe. A late medieval theologian attaches the idea of God's order to the fall of man as follows. By the Fall man was alienated from his true self. If he is to regain true self-knowledge he must do it through contemplating the works of nature of which he is part.

The poor wanderer, wishing to return to himself, should first consider the order of the things created by the Almighty; secondly he should compare or contrast himself with these; thirdly by this comparison he can attain to his real self and then to God, lord of all things.

Milton says just the same in *Of Education*, near the beginning:

The end then of learning is to repair the ruins of our first parents by regaining to know God aright. . . . But because our understanding cannot in this body found itself but on sensible things nor arrive so clearly to the knowledge of God and things invisible as by orderly conning over the visible and inferior creature, the same method is necessarily to be followed in all discreet teaching.

Later, in *Paradise Lost*, Milton makes Raphael carry out his educational plan. An 'orderly conning over the visible and inferior creature' forms the very first piece of instruction Raphael gives Adam on his visit to Paradise; to which Adam gratefully replies:

> Well hast thou taught the way that might direct
> Our knowledge, and the scale of Nature set
> From centre to circumference, whereon
> In contemplation of created things
> By steps we may ascend to God.

This double vision of world order and of the effects of sin was the great medieval achievement. Its origins, like those of the world order considered separately, go back to Genesis and Plato's *Timaeus* as brought together by the hellenizing Jews of Alexandria. Its great strength is that it admits of sufficient optimism and sufficient pessimism to satisfy the different tastes of varied types of man and the genius for inconsistency and contradiction that distinguishes the single human mind. Genesis asserts that when God had made the world he found it good and that he created man in his own image; but that with the Fall both man and (it was thence deduced) the universe also were corrupted. It was agreed that some vestiges of original virtue remained; but the proportions could be varied to taste, as Alice in Wonderland could nibble at both sides of the mushroom. The Platonic doctrine bore the same interpretation. The demiurge created the universe after the divine idea; hence the universe was good: but, as it was but a copy, it was removed from the idea and was thereby corrupted from perfection. As, further, both Plato and the orthodox Christian believed that man could rise above his imperfections and reach towards heavenly perfection, it was easy to identify the Platonic and Hebrew doctrines. How alive such an identification could be in the age of Elizabeth can be seen from Sidney, whose *Apology for Poetry* (for all its debt to Aristotle's *Poetics*) has it for its animating principle. More fundamental than any Aristotelian belief that poetry was more instructive than history or philosophy was the Neo-Platonic doctrine that poetry was man's effort to rise above his fallen self and to reach out towards perfection. As Sidney puts it:

Neither let it be deemed too saucy a comparison to balance the highest point of man's wit with the efficacy of nature, but rather give right honour to the Maker of that maker, who, having made man in his own likeness, set him beyond and over all the works of that second nature; which in nothing he showeth so much as in poetry, when with the force of a divine breath he bringeth things forth far surpassing her doings, with no small argument to the incredulous of that first accursed fall of Adam, sith our erected wit maketh us know what perfection is and yet our infected will keepeth us from reaching unto it.

Plato (or Plotinus) and Genesis are here amalgamated. The *divine breath* used to describe poetry and *perfection* derive from Plato; the references to Genesis are even more obvious. But we should think of both contexts together. The *perfection* is at once that of the Platonic Good and of the Garden of Eden, while Adam's fall from it is also the measure of the distance separating created things from their Platonic archetypes.

It was then through an intense realization of this double vision that the Elizabethans could combine such extremes of optimism and pessimism, about the order of the present world. The possibilities of great range were the greater because there was no tyranny of general opinion one way or another. This is one of the things that most separates the Elizabethan from the Victorian world. In the latter there was a general pressure of opinion in favour of the doctrine of progress: the pessimists were in opposition. In the Elizabethan world there was an equal pressure on both sides, and the same person could be simultaneously aware of each.

Finally, it is worth reflecting that *Paradise Lost* is in the main a traditional and orthodox poem. Though Milton held three major heresies, that the Son was not coeternal with the Father, that the soul died and was resurrected with the body, and that God created the world not out of nothing but out of himself, he does deal principally with the two great components of the traditional world picture: the glory of creation and the havoc sin made of it. The details of Old Testament history and of the life of Christ are subordinate to these. In spite of differences of date and of sympathies the theology of *Paradise Lost* is in its main lines a sound background of Elizabethan literature. It gets its emphasis in the right places.

In the foregoing accounts I may have been too insistent on the theological side. Anyone who has realized how general the opposite error has recently become can hardly avoid excess. When the most authoritative history of English literature calls Sir John Davies's *Nosce Teipsum* '*une sorte d'anomalie en ce temps de madrigaux et de pastorales*', it is difficult to speak moderately. How can anyone believing that begin to understand the Elizabethan age? Even E. K. Chambers in his welcome plea for a serious side to the Elizabethans does not go far enough. For it is insufficient to prove the seriousness of Daniel if the *Fairy Queen* is still counted pageantry and *Arcadia* a pastoral romance, the high philosophical rapture of the Garden of Adonis and the fierce Protestantism of Pamela in prison quite ignored. There has indeed been a mistaken trend to think of the Elizabethans as specialists in things secular or religious, as if no Elizabethan explorer could be a theologian, and no Londoner who heard a Puritan sermon ever saw a play. The instincts that send common humanity to see a bear-baiting or a prize fight, a popular play, or a popular preacher talking of death or hellfire, are all much the same: now and in the age of Elizabeth. Raleigh's life had been in part as secular as one can conceive: he must have felt and witnessed grossly human exultations and agonies in many parts of the globe: he must have known disorder at its most horrible both in nature and in man. Yet it is the same man who can see the glory of God not merely in abstraction from secular life, in the contemplation of the divine presence, but *per speculum creaturarum*.

In the glorious lights of heaven we perceive a shadow of his divine countenance; in his merciful provision for all that live his manifold goodness; and lastly, in creating and making existent the world universal, by the absolute art of his own Word, his power and almightiness; which power light virtue wisdom and goodness, being all but attributes of one simple essence and one God, we in all admire and in part discern *per speculum creaturarum*, that is, in the disposition order and variety of celestial and terrestrial bodies; terrestrial in their strange and manifold diversities; celestial in their beauty and magnitude, which in their continual and contrary motions are neither repugnant, intermixed, nor confounded. By these potent effects we approach to the knowledge of the omnipotent cause and by those motions their almighty mover.

4

THE CHAIN OF BEING

THE Elizabethans pictured the universal order under three main forms: a chain, a series of corresponding planes, and a dance. I shall deal with each picture in turn.

In a passage already referred to, Hooker spoke of all creation in these words:

> The general end of God's external working is the exercise of his most glorious and most abundant virtue. Which abundance doth show itself in variety, and for that cause this variety is oftentimes in Scripture expressed by the name of riches.

And here is Spenser's more explicit version of God's abundance in his *Hymn of Heavenly Beauty*.

> Then look, who list thy gazeful eye to feed
> With sight of that is fair, look on the frame
> Of this wide universe and therein read
> The endless kinds of creatures which by name
> Thou canst not count, much less their natures' aim;
> All which are made with wondrous wide respect
> And all with admirable beauty deckt.

Behind both passages is a traditional way of describing the world-order hinted at by Shakespeare in Ulysses' speech when he calls 'degree' the 'ladder to all high designs' and named by Pope in the *Essay on Man*, 'the vast chain of being'. It is the subject of a long and important book by Arthur Lovejoy. This metaphor served to express the unimaginable plenitude of God's creation, its unfaltering order, and its ultimate unity. The chain stretched from the foot of God's throne to the meanest of inanimate objects. Every speck of creation was a link in the chain, and every link except those at the two extremities was simultaneously bigger and smaller than another: there could be no gap. The precise magnitude of the chain raised metaphysical difficulties; but the safest opinion made it short of infinity though of a finitude quite outside man's imagination.

The idea began with Plato's *Timaeus*, was developed by Aristotle, was adopted by the Alexandrian Jews (there are signs of it in Philo), was spread by the Neo-Platonists, and from the Middle Ages till the eighteenth century was one of those accepted commonplaces, more often hinted at or taken for granted than set forth. The allegorizers interpreted the golden chain let down by Zeus from heaven in Homer as this chain of being. The eighteenth century inherited the idea of the chain of being, but, crassly trying to rationalize a glorious product of the imagination, ended by making it ridiculous and hence unacceptable in any form. Seeing the obvious chasm between men and even the lowest of the angels it insisted that there must be other inhabitable planets to house the intermediate orders of being. Thus when Pope in his *Essay on Criticism*, addressing the great poets of antiquity, says

> Nations unborn your mighty names shall sound,
> And worlds applaud that must not yet be found,

he means in the second line that these superior beings, to be discovered by the future flying-machines of the new science, will applaud Homer and Virgil.

It would be easy to accumulate texts describing the chain of being. One of the finest short accounts is by Sir John Fortescue, the fifteenth-century jurist, in his Latin work on the law of nature:

In this order hot things are in harmony with cold, dry with moist, heavy with light, great with little, high with low. In this order angel is set over angel, rank upon rank in the kingdom of heaven; man is set over man, beast over beast, bird over bird, and fish over fish, on the earth in the air and in the sea: so that there is no worm that crawls upon the ground, no bird that flies on high, no fish that swims in the depths, which the chain of this order does not bind in most harmonious concord. Hell alone, inhabited by none but sinners, asserts its claim to escape the embraces of this order. . . . God created as many different kinds of things as he did creatures, so that there is no creature which does not differ in some respect from all other creatures and by which it is in some respect superior or inferior to all the rest. So that from the highest angel down to the lowest of his kind there is absolutely not found an angel that has not a superior and inferior; nor from man down to the meanest worm is there any creature which is not in some respect superior to one creature and inferior to another. So that there is nothing which the bond of order does not embrace.

For a more detailed but quite popular account I know none better than that in a shortened version of the *Natural Theology* of Raymond de Sebonde. This was the work which Montaigne translated at his father's suggestion and which later he made the nominal theme of his most famous essay. The abbreviation was originally in Latin and was translated into French by Jean Martin in 1550. It is in the form of a dialogue between a priest and his pupil and is meant for the instruction of the young. As such it is commonplace in exactly the way that is wanted for evidence in this book. The account of the chain of being found here must have been the common property of western Europe in the sixteenth century. First there is mere existence, the inanimate class: the elements, liquids, and metals. But in spite of this common lack of life there is vast difference of virtue; water is nobler than earth, the ruby than the topaz, gold than brass: the links in the chain are there. Next there is existence and life, the vegetative class, where again the oak is nobler than the bramble. Next there is existence life and feeling, the sensitive class. In it there are three grades. First the creatures having touch but not hearing memory or movement. Such are shellfish and parasites on the base of trees. Then there are animals having touch memory and movement but not hearing, for instance ants. And finally there are the higher animals, horses and dogs etc., that have all these faculties. The three classes lead up to man, who has not only existence life and feeling, but understanding: he sums up in himself the total faculties of earthly phenomena. (For this reason he was called the little world or microcosm.) But as there had been an inanimate class, so to balance it there must be a purely rational or spiritual. These are the angels, linked to man by community of the understanding, but freed from simultaneous attachment to the lower faculties. There are vast numbers of angels and they are as precisely ordered along the chain of being as the elements or the metals. Now, although the creatures are assigned their precise place in the chain of being, there is at the same time the possibility of a change. The chain is also a ladder. The elements are alimental. There is a progression in the way the elements nourish plants, the fruits of plants beasts, and the flesh of beasts men. And this is all one with the tendency of man upwards towards God. The chain of being is educative both in the marvels of its static self and in its implications of ascent.

Such is the gist of Sebonde's account. Details can be added to it. There was for instance the question of transitions. If the chain is to be whole, the top of one class must link with the bottom of another. Here is Higden's account of these links from the second book of *Polychronicon*:

In the universal order of things the top of an inferior class touches the bottom of a superior; as for instance oysters, which, occupying as it were the lowest position in the class of animals, scarcely rise above the life of plants, because they cling to the earth without motion and possess the sense of touch alone. The upper surface of the earth is in contact with the lower surface of water; the highest part of the waters touches the lowest part of the air, and so by a ladder of ascent to the outermost sphere of the universe. So also the noblest entity in the category of bodies, the human body, when its humours are evenly balanced, touches the fringe of the next class above it, namely the human soul, which occupies the lowest rank in the spiritual order.

Among early Tudor translations from the Italian is one of the *Circe* of Gelli. The original, published in 1548, has for its theme a series of efforts by Ulysses to persuade some of his companions, now in bestial form through the magic of Circe, to return to their humanity. Circe is willing to change them back,

if they themselves wish it. The whole work is based on the idea of the chain of being, and the question of the beasts' metamorphosis is dependent on their position in it. Ulysses begins with the oyster (who had been a fisherman before his change), the lowest of the animals, who has of course the least chance of success. He goes up the scale of beasts with ever higher chances and in the end succeeds in persuading the king of the beasts, the elephant, to resume human form.

A charming attribute of the chain of being is that it allowed every class to excel in a single particular. The idea is Pythagorean or Platonic and it finds noble expression in the sixth chapter of the first book of Hooker's *Laws of Ecclesiastical Polity*. Stones may be lowly but they exceed the class above them, plants, in strength and durability. Plants, though without sense, excel in the faculty of assimilating nourishment. The beasts are stronger than man in physical energy and desires. Man excels the angels in his power of learning, for his very imperfection calls forth that power, while the angels as perfect beings have already acquired all the knowledge they are capable of holding. Only the angels, through their peculiar gift, the faculty of adoration, cannot claim to go beyond the class of being above them.

Another form of excellence, found in most accounts of the chain of being and certainly to be connected with it, is that within every class there was a primate. An example occurred above in Gelli's exaltation of the elephant. Sebonde speaks of the dolphin among the fishes, the eagle among birds, the lion among the beasts, the emperor among men. One of the most elaborate of these lists of primacy is in Peacham's *Complete Gentleman*, almost at the beginning. (It will be seen that opinion sometimes varied, the lion competing with the elephant, the whale with the dolphin, for primacy among beasts and fishes.)

If we consider arightly the frame of the whole universe and method of the all-excellent wisdom in her work as creating the forms of things infinitely divers, so according to dignity of essence and virtue in effect, we must acknowledge the same to hold a sovereignty and transcendent predominance as well of rule as place each over either. Among the heavenly bodies we see the nobler orbs and of greatest influence to be raised aloft, the less effectual depressed. Of elements the fire, the most pure and operative, to hold the highest place. The lion we say is king of beasts, the eagle chief of birds, the whale and whirlpool among fishes, Jupiter's oak the forest's king. Among flowers we most admire and esteem the rose, among fruit the pomeroy and queen-apple; among stones we value above all the diamond, metals gold and silver. And since we know to transfer their inward excellence and virtues to their species successively, shall we not acknowledge a nobility in man of greater perfection, of nobler form, and prince of these?

Other primacies were God among the angels, the sun among the stars, justice among the virtues, and the head among the body's members.

References to these primacies abound in literature but they lose greatly if it is not known that they are all part of a greater whole and that a reference to two or three implies both the rest of them and the ordered universe, in the background. For instance in *Richard II*, III, 3 Bolingbroke before Flint Castle says

> Be he the fire, I'll be the yielding water,

and a few lines later

> See, see, King Richard doth himself appear,
> As doth the blushing discontented sun
> From out the fiery portal of the east,
> When he perceives the envious clouds are bent
> To dim his glory and to stain the track
> Of his bright passage to the occident.

To which York adds,

> Yet looks he like a king: behold, his eye,
> As bright as is the eagle's, lightens forth
> Controlling majesty.

There in short space we have four of the traditional primacies:

fire among the elements, the sun among the planets, the king among men, the eagle among the birds. Again at the beginning of act five Richard is first a rose and then a lion.

As an aid to the poetic imagination the chain of being could work in several ways. First it made vivid the idea of a related universe where no part was superfluous; it enhanced the dignity of all creation, even of the meanest part of it. As one of the more pedestrian poets, Davies of Hereford, put it:

> The noblest creatures need the vil'st on ground,
> The vil'st are served by the honour'd most.
> And, which is more, the very heav'nly host
> Doth serve the basest creature void of sense
> Yet over-rules them in each clime and coast.
> So one to other have such reference
> As they in union have their residence.

And as one of the most poetical prose-writers put it:

Natura nihil agit frustra is the only indisputable axiom in philosophy. There are no grotesques in nature; nor anything framed to fill up empty cantons and unnecessary spaces.

Tennysonian doubts of nature's beneficence in preparing such excesses of seed over possible germination were excluded. The apparently superfluous could be put low down on the ladder of creation, and the matter was settled. Secondly, the chain was a metaphor that could help the mystically minded. Here was ultimate unity in almost infinite diversity and a means of ascent above the normal level of our nature. All the types in Spenser's Garden of Adonis are held fast to the base of God's throne. But it was the doctrine of plenitude that had the most obvious results in poetry; and here again the Garden of Adonis is one of the great illustrations.

> Infinite shapes of creatures there are bred
> And uncouth forms which none yet ever knew;
> And every sort is in a sundry bed
> Set by itself and rankt in comely rew:

Some made for beasts, some made for birds to wear;
And all the fruitful spawn of fishes' hew
In endless ranks along enranged were,
That seem'd the ocean could not contain them there.

Daily they grow and daily forth are sent
Into the world it to replenish more;
Yet is the stock not lessened nor spent
But still remains in everlasting store,
As it at first created was of yore.

He would be rash who conjectured what precise enthusiasms
had got themselves fixed in these wonderful lines: whether the
beginnings of new knowledge, the consciousness of national
unity, or the voyages of discovery, or even the personal
excitements of a man of genius feeling the teeming power of
his own mind. Whatever the content, social personal or
mystical, the idea of plenitude implied in the chain of being has
served its turn as an artistic instrument.

In his own way Milton exploits the notion of plenitude with
equal success. Even if, according to Lovejoy, the 'dialectic of
the idea of plenitude' had little to do with 'determining his
scheme of things', his living sense of plenitude had much with
determining his poetry. Milton's forthright exposition of the
chain of being is Raphael's first piece of instruction to Adam on
his visit to Eden. (With superb cunning Milton calls Raphael
'the winged Hierarch', to summon up in a word the associa-
tions of 'degree'.)

To whom the winged Hierarch repli'd:
O *Adam*, one Almightie is, from whom
All things proceed, and up to him return,
If not deprav'd from good, created all
Such to perfection, one first matter all
Indu'd with various forms, various degrees
Of substance, and in things that live, of life;
But more refin'd, more spiritous, and pure,
As neerer to him plac't or neerer tending
Each in thir several active Sphears assigned,

> Till body up to spirit work, in bounds
> Proportioned to each kind. So from the root
> Springs lighter the green stalk, from thence the leaves
> More aerie, last the bright consummate floure
> Spirits odorous breathes: flours and thir fruit
> Mans nourishment, by gradual scale sublim'd
> To vital Spirits aspire, to animal,
> To intellectual, give both life and sense,
> Fansie and understanding, whence the soul
> Reason receives, and reason is her being,
> Discursive, or Intuitive; discourse
> Is often yours, the latter most is ours,
> Differing but in degree, of kind the same.

Milton insists too on the abundance of angelic creatures and their many functions when he makes Adam say to Eve:

> Millions of spiritual Creatures walk the Earth
> Unseen, both when we wake, and when we sleep:
> All these with ceaseless praise his works behold
> Both day and night.

But poetically his most powerful expositions of the doctrine of plenitude are elsewhere. First there is Comus's speech in temptation of the Lady. Comus is here a formidable foe. He is the ape of God, repeating but distorting the cosmic ecstasies of the theologians:

> Wherefore did Nature powre her bounties forth,
> With such a full and unwithdrawing hand,
> Covering the earth with odours, fruits, and flocks,
> Thronging the Seas with spawn innumerable –

So far he is the orthodox praiser of God's bounty, but when he adds

> But all to please, and sate the curious taste?

he declares himself the diabolic epicurean, the blasphemer of the bounty of God. But the poetry and the sense of nature's exuberance are not impaired. Then there is the riot of growth in the Garden of Eden, the 'enormous bliss', in the fourth book of *Paradise Lost*, and finally the whole description of creation in book seven culminating in man.

The other major work that demands mention here is the *Tempest*. With the general notion of order Shakespeare was always concerned, with man's position on the chain of being between beast and angel acutely during this tragic period; but only in the *Tempest* does he seem to consider the chain itself. Here indeed man is distanced into a more generally cosmic setting. The heavens are actively alive. It was by Providence divine that Prospero and Miranda survived in the boats. Destiny has this lower world as its instrument. The thunder proclaims Alonso's guilt. Ariel and the other angelic powers, elves and demi-puppets, are all, according to W. C. Curry, in the orthodox tradition of the Renaissance Neo-Platonists, who were the great exponents of the chain of being. Prospero is at the apex of humanity with his magic power and his decision to spend what remains of his life in contemplation. Trinculo and Stephano are low in the scale of humanity. Caliban is largely bestial, a better log-carrier than a man and perhaps of cruder appetites, strong too in fancy in which according to one Renaissance theory beast excelled man. Nor are the beasts forgotten. Prospero tells Ariel that his groans from the pine-tree prison

> Did make wolves howl and penetrate the breasts
> Of ever angry bears.

The whole play is alive with the sense of creation's flux and not blind to creation's limit. Caliban may hover between man and beast, yet in the end he shows himself incapable of the human power of education. Prospero too learns his own lesson. He cannot transcend the terms of his humanity. In the end he acknowledges Caliban, 'this thing of darkness, mine': man for all his striving towards the angels can never be quit utterly of the bestial, of the Caliban, within him.

It is interesting that Shakespeare was apt to use the traditional lore, described in this chapter, at high moments, just as he used the traditional comparison between the body politic and the microcosm to picture worthily the terrible struggle in the mind of Brutus. The references quoted from *Richard II* to leaders in different divisions of creation (sun, rose, etc.) did not indeed evoke the highest poetry, but take these two passages from *Antony and Cleopatra* and *Coriolanus* respectively. Cleopatra, praising Antony to Dolabella in a speech full of cosmic references, says

> his delights
> Were dolphin-like; they show'd his back above
> The element they lived in.

The passage loses half its meaning unless the reference to the dolphin as king of the fish is understood. Antony stood out in regal fashion above the revels he delighted in like the dolphin, king of the fishes, showing his back above the waves. Aufidius, speculating how Coriolanus will deal with Rome, says:

> I think he'll be to Rome
> As is the aspray to the fish, who takes it
> By sovereignty of nature.

The aspray, or osprey, was a small eagle, king among birds, and fish were supposed to yield themselves voluntarily, turning their bellies up to him. Coriolanus has been gifted from birth with the regal touch. It is the same kingliness that is attributed (and probably by Shakespeare) to Theseus in the first scene of the *Two Noble Kinsmen*

> Remember that your fame
> Knolls in the ear o' th' world: what you do quickly
> Is not done rashly; your first thought is more
> Than others' laboured meditance; your premeditating
> More than their actions: but, Oh Jove, your actions,
> Soon as they move, as asprays do the fish,
> Subdue before they touch.

The authors from whom I have so far quoted in this chapter

can on the whole be called optimists. In spite of original sin and the corruption it imparted to the natural world, God's great plan still stood out conspicuous in his works. But there were those who thought the whole creation corrupt and in its decrepitude. Donne (in some moods) was one. Another is Christopher Goodman, who published in 1616 a book called the *Fall of Man*. He was answered by George Hakewill in the *Power and Providence of God in the Government of the World*, which maintains that nature is still bright as on creation's day. This controversy is wider known than it might be because Milton (taking Hakewill's side) contributed remotely to it in a Latin poem written at college called *Naturam non pati senium*. What vitality the idea of the chain of being had and the lesson it taught is shown by Goodman's being unable to ignore it. In spite of nature's corruption he has to admit the educative force of the *speculum creaturarum*. The chain of being is still a means of spiritual ascent: but only in an ideal interpretation. God has a ceremonial as well as a natural law, and it is possible to ascend spiritually by looking on nature ceremonially:

This I speak, supposing there were some joy in the creature, which, if there were any, thy mind might be transported and carried by the ladder or bridge of the creatures to the love of thy creator. For as it pleased God to ordain a ceremonial law differing from the natural law, according to the wisdom of his own institution, so assuredly the mind of man, which delights in nothing so much as in mysteries, may make whole nature a ceremony and all the creatures types and resemblances of spiritual things.

5

THE LINKS IN THE CHAIN

I. Angels and Ether

IT will be convenient to describe the Elizabethan scheme of creation from top to bottom. But first we must shed any notion that even in the Middle Ages the chain or ladder of creation was single and consistent. Some portions of it plainly could not be fitted into a single unit: for instance the four elements. These, as inanimate, should ideally be lower than the lowest animate creation. It might have been expected that fire, the highest of the elements, should link with the vital spark of a worm or an oyster. But the operations of the elements did not cease with the lowest living thing: nor were the higher living things compounded of the lower, but all were compounded of the four elements direct. So the elements could not be links in a simple chain: they had to be a supplementary chain multifariously connected with the main one. Now just as the elements, themselves inanimate, touched the chain at higher or at least middle as well as at low places, so the perfected parts of inanimate nature were arranged in a hierarchy that removed its heights far above the lower specimens of the animate class. In other words the upper reaches of the physical universe were connected not with plants or beasts (though they might act on them) but with the very angels, with whom they will have to be considered.

In spite of Copernicus and a wide knowledge of his theories through popular handbooks, the ordinary educated Elizabethan thought of the universe as geocentric. He was as apt as a modern to meditate on its immensity and he thought of God as domiciled beyond the bounds of the fixed stars in the *coelum empyraeum* (Milton's *empyrean*) attended by the hosts of the angels. The name of this heaven signifies fire and hence, fire being the best of the elements and heavenly fire being better than the elemental, the highest perfection. It also signifies light and the notion of God as light. Milton's first description of heaven cannot be far from the Elizabethan conception:

> Now had the Almighty Father from above,
> From the pure Empyrean where he sits
> High Thron'd above all highth, bent down his eye,
> His own works and their works at once to view:
> About him all the Sanctities of Heaven
> Stood thick as Starrs, and from his sight receiv'd
> Beatitude past utterance.

What if anything there was between the *coelum empyraeum* and the created universe was not settled, but on one notion there was an intermediate space containing lesser heavens. To what use it was put I will return when I speak in detail of the angels. Opinion varied on the precise constitution of the created universe. The number of spheres that composed it could be nine ten or eleven; but no one doubted that round a central earth revolved with differing motions spheres of diameters ever increasing from the moon's through the other planets to that of the fixed stars, and that there was a sphere called the *primum mobile* outside that of the fixed stars, which dictated the motions proper to all the rest. Within this universe there was a sharp division between everything beneath the sphere of the moon and all the rest of the universe. (The adjective *sublunary* contains a lot of meaning.) It was the difference between mutability and constancy. Though the four elements were the material for the whole universe, they were differently mixed in these two regions: below the moon ill, above it perfectly. Hence the heavens were eternal, the sublunary regions subject to decay: on the medieval principle that, in Donne's words, whatever dies was not mixed equally. Another difference was that, while below the moon the air was thick and dirty, above it was pure and known as the ether. In the words of an encyclopedist printed by Caxton:

This air shineth night and day of resplendour perpetual and is so clear and shining that if a man were abiding in that part he should see all, one thing and another and all that is, fro one end to the other, all so lightly or more as a man should do here beneath upon the earth the only length of a foot or less.

And there was an alternative theory which made the ether a fifth element and the substance of all creation from the moon upward. As was only natural, the farther the distance from the earth and the nearer to heaven, the purer and more brilliant was the atmosphere. Contrariwise the earth itself was gross and heavy and the more so towards its own centre. Far from being dignified and tending to an insolent anthropocentricity, the earth in the Ptolemaic system was the cesspool of the universe, the repository of its grossest dregs. As a Frenchman of the time of Francis I expressed it, the earth

is so depraved and broken in all kind of vices and abominations that it seemeth to be a place that hath received all the filthiness and purgings of all other worlds and ages.

Nor did the Ptolemaic system make for any sense of smallness or confinement. It was just as possible in Elizabethan days as later to be terrified by the vast spaces. Caxton's encyclopedist must have taken away the breath of the vulgar quite as effectively as any modern lecturer on the marvels of the heavens; and he uses the same method. This is how he puts the immense distance of the stars from the earth:

If the first man that God formed ever, which was Adam, had gone fro the first day that he was made and created twenty-five miles every day, yet should he not have comen thither, but should yet have the space of seven hundred and thirteen year to go at the time when this volume was performed by the very author. Or if there were a great stone which should fall fro thence unto the earth it should be an hundred year ere it came to the ground.

It has been necessary to describe the physical universe down to the very earth because the angels (as a species though not individually) could inhabit or visit its whole range. As in so many other matters, the Elizabethans kept the main medieval beliefs about the angels but omitted or confused many of the details. First they were convinced that there were angels and would have agreed with Sir Thomas Browne that it is a riddle

how so many learned heads should so far forget their metaphysic and destroy the ladder and scale of creatures as to question the existence of spirits.

Further they were quite clear that angels are intermediate between God and man; that their nature is purely intellectual; that they possess free will like man, but that it never conflicts with God's will; that they can apprehend God immediately and not by figure or symbol; that they are arranged in orders; that they are God's messengers; and that they act as guardians of men. All this the Elizabethans held in common with the Middle Ages. But just as the reformers abolished much of the church ceremony, so the Elizabethans ignored many attributes of the angelic hierarchies.

As before, Sebonde gives an admirable general account of the conventional medieval notion of the orders of the angels:

We must believe that the angels are there in marvellous and inconceivable numbers, because the honour of a king consists in the great crowd of his vassals, while his disgrace or shame consists in their paucity. Thousands of thousands wait on the divine majesty and tenfold hundreds of millions join in his worship. Further, if in material nature there are numberless kinds of stones herbs trees fishes birds four-footed beasts and above these an infinitude of men, it must be said likewise that there are many kinds of angels. But remember that one must not conceive of their multitude as confused; on the contrary, among these spirits a lovely order is exquisitely maintained.

Sebonde goes on to speak of the various orders on earth and adds:

If then there is maintained such an order among low and earthly things, the force of reason makes it necessary that among these most noble spirits there should be a marshalling unique, artistic, and beyond measure blessed. Further, beyond doubt, they are divided into three hierarchies or sacred principalities, in each of which there are high, middle, and low.

The most influential account of the angels had been that of the man traditionally called Dionysius the Areopagite, a Christian Neo-Platonist of the fifth century A.D., in his work *On the Heavenly Hierarchy*; and it is widely known because Aquinas and Dante accepted it. Dionysius taught that the angels are arranged in a definitive order according to their natural capacity to receive the undivided divine essence. Knowing themselves and being without sin, they are utterly content with the full measure of what they can assimilate, and will not envy those above them. Those of inferior capacity will receive divine knowledge through the medium of their superiors. There are three main orders of angels. The highest is contemplative and consists of Seraphs, Cherubs, and Thrones. Thus the highest link in the chain of being would be the chief Seraph. The second order is more active but rather potentially than in deed: their psychological state is rather of an attitude than of an action. They are divided into Dominations, Virtues, and Powers. More active still is the third order, divided into Principalities, Archangels, and Angels. It is this lowest rank, the Angels, who form the medium between the whole angelic hierarchy and man. They go on God's errands. To the medieval mind the nine orders mattered a great deal, first because the triple divisions echoed the Trinity, second because they corresponded to the ninefold division of the material heavens, accepted in the main in medieval times. From heaven in descending order the spheres were those of *primum mobile*, the fixed stars, Saturn, Jupiter, Mars, the Sun, Venus, Mercury, and the Moon; and the nine hierarchies of angels were thought each to regulate one of these spheres in the order given above, the seraphs regulating the *primum mobile* and so on. Dionysius also sees a correspondence between the angelic hierarchy in heaven and the ecclesiastical hierarchy on earth.

The Dionysian order was not unknown through the Elizabethan age. But it had lost its authority, and was sometimes ignored and sometimes altered. It is worth here asking the question (which I have not seen asked before) whether the best known of all Elizabethan references to an order of angels may not be an accurate rendering of part of the Dionysian plan. In the *Merchant of Venice* Lorenzo says to Jessica:

> There's not the smallest orb which thou behold'st
> But in his motion like an angel sings,
> Still quiring to the young-eyed Cherubins.

These lines are usually explained by the Platonic doctrine of the music of the spheres. But in Plato each sphere made its own note; there was no question of all the stars singing. Shakespeare imagines all the stars in the sphere of the fixed bodies singing to the Cherubim. Now in the Dionysian scheme it is the Cherubim who have charge of the fixed stars. Was it a mere accident that Shakespeare wrote *Cherubin* rather than *Seraphin* (euphony apart), or did he know the tradition?

For a knowledge of medieval doctrine but a free adaptation of it one can go to Milton. He has his various hierarchies but lays down no precise order and diverges from Dionysius by exalting the Archangels to the supreme place, above the Seraphim and Cherubim. Donne is familiar with many of the medieval details and in using them assumes a like familiarity in his audience. His *Air and Angels* has as its main point the doctrine that the angels are of a brightness insufferable to human sight and that when they appear to men they assume a body from the ether:

> For nor in nothing nor in things
> Extreme and scatt'ring bright can love inhere;
> Then as an angel face and wings
> Of air, not pure as it, yet pure doth wear,
> So thy love may be my love's sphere.

The angel lore here is precise and abundant. The comparison is between love and an angel finding embodiment. An angel takes his body neither from nothing nor from the fiery element of the empyrean. He is himself purely spiritual and for embodiment chooses something grosser than himself, yet not unworthy of his own purity. This is the ether, the pure air surrounding the heavenly spheres.

Once more Hooker probably gives the average beliefs of the well-educated Elizabethan. Angels, according to him, are

spirits immaterial and intellectual, the glorious inhabitants of those sacred palaces, where nothing but light and blessed immortality, no shadow of matter for tears discontentments griefs and uncomfortable passions to work upon, but all joy tranquillity and peace even for ever and ever doth dwell: as in number and order they are huge mighty and royal armies, so likewise in perfection of obedience unto that law which the Highest, whom they adore love and imitate, hath imposed upon them . . . God, which moveth mere natural agents as an efficient only, doth otherwise move intellectual creatures and especially his holy angels: for beholding the face of God, in admiration of so great excellency, they all adore him; and being rapt with the love of his beauty they cleave inseparably for ever unto him. Desire to resemble him in goodness maketh them unweariable and even unsatiable in their longing to do by all means all manner good unto all the creatures of God, but especially unto the children of men: in the countenance of whose nature, looking downward, they behold themselves beneath themselves; even as upward in God, beneath whom themselves are, they see that character which is nowhere but in themselves and us resembled.

Hooker does not wish to arouse scholastic controversy and keeps to what were for his age essentials: the nature of angelic intelligence, their place in the scale of being, and their function as guardians of men.

The notion of the Guardian Angel was universal and is constantly found in the poets. The Attendant Spirit in *Comus* is quite orthodox, though put in a quasi-pastoral setting:

> Before the starry threshold of *Joves* Court
> My mansion is, where those immortal shapes
> Of bright aërial Spirits live insphear'd
> In regions milde of calm and serene Ayr,

Above the smoak and stirr of this dim spot,
Which men call Earth, and with low-thoughted care
Confin'd, and pester'd in this pin-fold here,
Strive to keep up a frail, and Feaverish being ...

The spirit is an intellectual being living above the sublunary
realm, the terrestial cesspool, here likened to a cattle-pen, in the
ether, in the region of the incorruptible spheres on the borders
of God's court, the Empyrean. And these two stanzas from the
Fairy Queen are central in Elizabethan belief:

And is there care in heaven? And is there love
In heavenly spirits to these creatures base,
That may compassion of their evils move?
There is: else much more wretched were the case
Of men than beasts. But oh th' exceeding grace
Of highest God that loves his creatures so
And all his work with mercy doth embrace,
That blessed angels he sends to and fro
To serve to wicked men, to serve his wicked foe.

How oft do they their silver bowers leave
To come to succour us that succour want;
How oft do they with golden pinions cleave
The flitting skies like flying pursuivant
Against foul fiends to aid us militant.
They for us fight, they watch and duly ward
And their bright squadrons round about us plant;
And all for love and nothing for reward.
Oh, why should heavenly love to men have such regard?

However, though Hooker may speak for the majority, there
are other beliefs about angels too persistent in the age of
Elizabeth to be omitted. These are mainly derived from the
renewed cult of Plato and Plotinus that began with Ficino, the
Florentine translator and annotator of both philosophers, and
was popularized in England through various channels. One of
the chief of these was the discourse of Bembo on love in the
last book of Castiglione's *Courtier*, widely known through

Hoby's translation in 1566. This renewed Platonizing created an enthusiastic idealism which is a true mark of the Renaissance, that phase of culture which nowadays tends ever more to lose its identity and to turn out to be simply the late Middle Ages. It is a habit of mind most difficult for a modern to grasp, being at once fantastic and closely allied to action. It was something that impelled Sidney to seek education through his love for Stella, and honour in sordid battles in the Low Countries; that turned Queen Elizabeth into Belphoebe without in the least blunting men's knowledge that she was a difficult and tyrannical old woman. In the same way it fostered a high and fantastical conception of the universe among men who lived in an England whose standards of hygiene decency and humanitarianism would make a modern sick.

It was mainly the Platonists who made less tidy and more picturesque a world which in the medieval view had been given a mathematical neatness. Caxton's encyclopedist is perfectly clear and neat in the following passage:

> God formed the world all round like as is a pellet the which is all round; and he made the heaven all round which environeth and goeth round about the earth on all parts wholly without any default, all in like wise as the shell of an egg that environeth the white all about. And so the heaven goeth round about an air which is above the air, the which in Latin is called *hester* [*aether*].

And immediately outside the egg is God's heaven. Though there *had* been medieval Platonists, this quotation gives the prevalent tone. Set against this a piece of Italian Platonizing, a brief description of the universe from Count Hannibal Romei's *Courtier's Academy* translated into English from the Italian by I. K. in 1598. Romei in the usual manner ascends the ladder of creation but above man and below the angels he puts Nature as an intellectual being. She is Plato's 'Soul of the world' and as such gives life and shape to it. As nature, she 'imprinteth into matter with the seal of divinity all forms generative and corruptible'. As Intelligence, she is unerring and 'directeth everything deprived of understanding to their end'. Above Nature there follow the angels arranged in the Dionysian order.

This giving a soul to nature – nature, that is, in the sense of *natura naturans*, the creative force, not of *natura naturata*, the natural creation – was a mildly unorthodox addition to the spiritual or intellectual beings; and I had better give the orthodox opinion at this point. For the Elizabethans talked much about nature, and she cannot be omitted from the world picture. That there was a law of nature was universally agreed; she worked unswervingly by a set of rules applicable to her alone; but the question still remained whether she was a voluntary or involuntary agent. Hooker, orthodox as usual, is explicit on this matter. She cannot be allowed a will of her own or the rank of a kind of goddess. She is not even an agent with her eye ever fixed on God's principles; rather she is the direct and involuntary tool of God himself. The different phenomena of nature *must* perform their own proper functions in order to retain their identities, but they are not conscious of this. They work not on nature's initiative but on God's, by the pressure of his Providence. 'Nature is nothing but the tool of God.' Hakewill applies the same doctrine to the various heavenly bodies. When the Psalmist speaks of the sun as knowing his going down he speaks but figuratively:

the Prophet thereby implying that the sun observeth his prescribed motion so precisely to a point that in the least jot he never erreth from it. And therefore is he said to do the same upon knowledge and understanding; not that the sun hath any soul or use of understanding but because it keepeth his courses and measures exactly according to God's prescription . . . That argument from whence the heathen do collect that the stars must needs be gods doth most plainly prove the contrary. For if they take them to be gods because of the certainty of their courses they be therein much deceived; for this plainly proveth that indeed they be no gods because they be not able to depart from their set courses. Whereas, if they were gods, they would move both this way and that way in the heavens as freely as living creatures do upon the earth, who, because they have the liberty and freedom of their will, they wander up and down whither they themselves please.

Another piece of Platonizing, beautiful but irresponsible, is found in Spenser, in his *Hymn of Heavenly Beauty*. Here the spheres of the physical universe are not regulated by the different orders of angels on the medieval scheme but duplicated Platonically by other, ideal ones:

> For far above these heavens, which here we see,
> Be others far exceeding these in light,
> Not bounded, not corrupt, as these same be,
> But infinite in largeness and in height,
> Unmoving, uncorrupt, and spotless bright,
> That need no sun t'illuminate their spheres
> But their own native light far passing theirs.
>
> And as these heavens still by degrees arise
> Until they come to their first mover's bound,
> That in his mighty compass doth comprise
> And carry all the rest with him around;
> So those likewise do by degrees redound
> And rise more fair, till they at last arrive
> To the most fair, whereto they all do strive.

In these heavenly spheres Spenser sets a strange variety of inhabitants: in the lowest, the souls of the righteous men; in the next, Plato's Ideas and Intelligences; in the next, Powers and Potentates; in the next, Seats and Dominations; in the next, Cherubim and Seraphim; and last, attending on God himself, the Angels and Archangels. That it should really *matter* to Spenser that he should insert Platonic ideas into the order of heaven is a measure of the queerness of the Elizabethan age.

Another powerful and persistent piece of Platonism has to do with the angels who were supposed to direct the turning of the spheres in the physical universe. This notion, in itself Platonic, is best known to the common reader today through the references in the poems of Donne, as when at the beginning of *Good Friday: Riding Westward* he compares the soul to a sphere and devotion to the Intelligence or Angel revolving it. This in Donne is not just a piece of private medievalism but a current orthodox notion. Hooker hints at it. Goodman finds the motion of the spheres so complicated that it must be directed by Intelligences, 'which in effect are angels'. Even the sphere of the moon is mysterious, as it has the strange power of drawing the tides. Hakewill, Goodman's opponent, accepts the Intelligences by tradition:

It is the joint consent of the Platonics Peripatetics and Stoics, and of all the noted sects of philosophers who acknowledged the divine providence, with whom agree the greatest part of our most learned and Christian doctors, that the heavens are moved by angels.

Nor were the Intelligences unimportant to the theory of the chain of being. Before matters were complicated by the notion of other worlds peopled by intermediate creatures, the Intelligences were often thought of as the lowest rank of angels linking with the highest specimens of the human race.

Where the Platonists found delight was in elaborating the function of the Intelligences. Not only, they maintained, did angels move the spheres, but these angels were identical with those heavenly sirens who, in Plato, sit upon their spheres and, each singing their different note, compose a harmony of ravishing beauty. Aristotle had laughed at the notion, which, with the two masters on different sides, became a stock theme of academic disputation. Hakewill puts the matter conveniently thus:

There were among the Ancients not a few, nor they unlearned, who by a strong fancy conceived in themselves an excellent melody made up by the motion of the celestial spheres. It was broached by Pythagoras, entertained by Plato, stiffly maintained by Macrobius and some Christians such as Beda, Boethius, and Anselmus of Canterbury; but Aristotle puts it off with a jest as being a pleasant and musical conceit but in effect impossible.

Whatever the issue of the dispute, the notion of the spherical music fascinated men's minds and inspired the poets. The

Genius of the Wood in Milton's *Arcades* describes how he listens to the heavenly music when his duties as forest-guard are done for the day:

> But els in deep of night when drowsines
> Hath lockt up mortal sense, then listen I
> To the celestial *Sirens* harmony,
> That sit upon the nine enfolded Sphears,
> And sing to those that hold the vital shears,
> And turn the Adamantine spindle round,
> On which the fate of gods and men is wound.
> Such sweet compulsion doth in music ly,
> To lull the daughters of *Necessity*,
> And keep unsteddy Nature to her law,
> And the low world in measur'd motion draw
> After the heavenly tune, which none can hear
> Of human mould with grosse unpurged ear.

This last Platonic addition, that the grossness of the flesh stops us hearing the music of the heavens or rather its musical counterpart in our own microcosms, was known also to Shakespeare. Lorenzo says to Jessica after referring to the quiring orbs:

> Such harmony is in immortal souls;
> But whilst this muddy vesture of decay
> Doth grossly close it in, we cannot hear it.

A further imagining – perfect example of that fusion of Plato and Genesis that appealed as strongly to the Elizabethans as to any generation that had accepted it – was that before the Fall man *could* hear the music. It is of a piece with the passage from Sidney's *Apology* I have referred to already. The poet's wit reaches beyond actual fact, for example to the conception of the music of the spheres,

with no small argument to the incredulous of that first accursed fall of Adam, sith our erected wit maketh us know what perfection is and yet our infected will keepeth us from reaching unto it.

If the Elizabethans were convinced that there were realms of

purity and bliss above the sublunary sphere, that angels of
different kinds inhabited them, and that some of these angels
did God's errands or protected men, they were equally convinced that a part of the angels fell from grace, inhabited hell,
and did harm to men. There was agreement in every sect that
they fell through pride. Hooker on this topic is both traditional
and representative of current opinion. The bad angels fell
away voluntarily, and they did so because they turned their
minds away from God and from God's creation, itself the
evidence of God's goodness, to themselves. There was indeed

no other way for angels to sin but by reflex of their understanding upon
themselves; when, being held with admiration of their own sublimity
and honour, the memory of their subordination unto God and their
dependency on him was drowned in this conceit. Whereupon their
adoration love and imitation of God could not choose but be also
interrupted. The fall of the angels was therefore pride.

Further, it was generally agreed, according to the early Christian doctrine, that the fallen angels took on the form of Pagan
deities or dispersed themselves into various parts of the physical
universe. Hooker speaks of them as

dispersed, some in the air, some on the earth, some in the water, some
among the minerals dens and caves that are under the earth. . . . These
wicked spirits the heathen honoured instead of gods, both generally
under the name of gods infernal and particularly some in oracles, some
in idols, some as household gods, some as nymphs: in a word, no foul
and wicked spirit which was not one way or other honoured of men
as God, till such time as light appeared in the world and dissolved the
works of the Devil.

But the lore of angels and devils was so fascinating that it
lingered on in some quarters with much of its medieval detail.
Burton's long chapter on the nature of spirits is the classic
exposition, and it is worth quoting from him an example of
what the Middle Ages had taken so earnestly but was degraded
to fascinating antiquarianism in the age of Elizabeth.

Our schoolmen and other divines make nine kind of bad spirits, as

Dionysius hath done of angels. In the first rank are those false gods of the gentiles, which were adored heretofore in several idols and gave oracles at Delphos and elsewhere, whose prince is Beelzebub. The second rank is of liars and equivocators, as Apollo Pythius and the like. The third are those vessels of anger, inventors of all mischief; their prince is Belial. The fourth are malicious revenging devils, and their prince is Asmodeus. The fifth kind are cozeners, such as belong to magicians and witches; their prince is Satan. The sixth are those aerial devils that corrupt the air and cause plagues thunders fires etc., spoken of in the Apocalypse, and Paul to the Ephesians names them the princes of the air; Meresin is their prince. The seventh is a destroyer, captain of the Furies, causing wars tumults combustions uproars, mentioned in the Apocalypse and called Abaddon. The eighth is that accusing or calumniating devil, that drives men to despair. The ninth are those tempters in several kinds, and their prince is Mammon.

To the mathematically minded Middle Ages it really mattered that there should be nine hierarchies of bad angels to match the nine of the good; but Burton goes on casually to say that Psellus divides devils into six. Not that the belief in devils and witches was not very much alive in Burton's day: only it was held differently.

The passage from Burton has brought us from the mystical rapture of Dionysius's sublime heaven to the homelier tangles of folklore and superstition. Here orthodox angels merge with fairies, and devils with pixies and succubi. It is a vast area that must lie outside the scope of the present book.

The reader may think that in a book that deals with the permanent assumptions of the Elizabethans I have made too much of the angels. Was it really the case, he may ask, that they were there, all the time, as part of the general scheme of things? It seems as if they were and that men were really conscious of them as above themselves in virtue and yet not unlike. These words of Hooker probably represent the common assumptions:

Neither are the angels themselves so far severed from us in their kind and manner of working but that between the law of their heavenly operations and the actions of men in this our state of mortality such correspondence there is as maketh it expedient to know in some sort the one for the other's more perfect direction.

II. *The Stars and Fortune*

It has been well said that for the Elizabethans the moving forces of history were Providence, fortune, and human character. The present survey of the chain of being pretty much follows this course. So far the theme has been the divine order and God's pressure in maintaining it, the changeless heavens and the perfected spiritual beings. We now come to the stars which, through obeying God's changeless order, are responsible for the vagaries of fortune in the realms below the moon. The planets were in fact the commuting agents of eternity to mutability: they had the function of the million pieces of coloured glass which in Shelley's dome stain the white radiance of eternity, when the glass itself does not change but causes change in something else.

Though the images used to express the sway of the stars and of fortune are different, the two influences on the world are the same. For the sway of fortune the image of the wheel is constant both in literature and in picture; and at times it is presented with such concrete circumstance as both to risk absurdity and to turn the spectator's or the reader's thoughts far from the stars with their subtly penetrating influences. There are those grossly physical pictures of human beings, realistically dressed, clinging or tied to what seems a large cart wheel, in process of being turned aloft or hurled in undignified somersault onto the ground; or there is this from Hamlet:

> Out, out, thou strumpet, Fortune. All you gods,
> In general synod take away her power;
> Break all the spokes and fellies from her wheel
> And bowl the round nave down the hill of heaven
> As low as to the fiends.

Nevertheless it was quite taken for granted that the stars

dictated the general mutability of sublunary things, and that fortune was part of this mutability applying to mankind alone.

At the time when Christianity was young and growing, there was general terror of the stars and a wide practice of astrology. The terror was mainly superstitious, and the only way of mitigating the stars' enmity was through magic. It was one of the Church's main tasks to reduce the licence of late pagan astrological superstition to her own discipline. There was no question of cutting it out altogether. Naturally she did not wholly succeed and her task could never be completed. In the Elizabethan as in earlier ages the orthodox belief in the stars' influence, sanctioned but articulated and controlled by the authority of religion, was not always kept pure from the terrors of primitive superstition.

With the superstitious terrors I am not concerned: they have little specifically to do with the Elizabethan age. But it is worth reflecting (as is not always done) that even these were not all horror and loss. If mankind had to choose between a universe that ignored him and one that noticed him to do him harm, it might well choose the second. Our own age need not begin congratulating itself on its freedom from superstition till it defeats a more dangerous temptation to despair. Neither am I concerned with the details of practical astrology. Though horoscopes were cast and much minute doctrine survived, the rules of the game, as I pointed out above, had changed. Indeed the difference between medieval and Elizabethan astrological belief and practice is rather like that between real tennis and its simplified progeny. But both games were played with zest.

Whether he were scrupulously orthodox or inclined to instinctive superstition, the Elizabethan believed in the pervasive operation of an external fate in the world. The twelve signs of the zodiac had their own active properties. The planets were busy the whole time; and their fluctuating conjunctions produced a seemingly chaotic succession of conditions, theoretically predictable but in practice almost wholly beyond the wit of man. Their functions differed, with the moon the great promoter of change. Though there were sceptics like Edmund in *Lear* and though the quack astrologer was hated and satirized, the general trend was of belief.

It must not be thought that the evident havoc in nature's order wrought by the stars at all upset the evidence of God's Providence. The havoc was all within the scheme. The answer to the question why God allowed the havoc was almost self-evident. It was not primarily God who allowed it but man who inflicted it on both himself and the physical universe. In their own natures the stars are beneficent, and when they were first created they worked together to do good. In Goodman's words:

The stars in general intend the earth's fruitfulness; each one in particular hath his several office and duty: if virtue be added to virtue and their influence together concur, it should rather further and perfect the action. Certainly some overruling hand and providence stirs up these uproars and thereby intimates the reciprocal opposition, as of the earth to the heavens so of the heavens to the earth. But the root of the dissension first bred and is still fastened to the earth, from whence proceeds the first occasion of these tumults.

And later Goodman narrows still further the origin of the trouble:

Now in this great uproar and tumult of nature, when heaven and earth seem to threaten a final destruction, give me leave with the mariners of Jonah's ship to cast lots and search out the first occasion of this evil. Alas, alas, the lot falls upon man: man alone of all other creatures, in regard of the freedom of his will and the choice of his own action, being only capable of the transgression, the rest of the creatures are wholly excluded from the offence; the punishment (I confess) appears in them but chiefly and principally in man.

It was the Fall, then, that was primarily responsible for the tyranny of fortune, and, this being so, man could not shift the blame but must bear his punishment as he can. It was God of course who, prompted by the Fall, set the celestial bodies against each other in their influence on the sublunary universe, but he also tempers their opposition as a prudent king sets one ambitious noble against another, thereby preserving a balance of power. As a certain John Norden in a dull poem on the vicissitude of things put it:

> Yet thus this disagreement must be set
> As in the discord be no power to wrong.
> For why? supremest have no fatal let
> But will prevail, as they become too strong.
> Therefore such mean must them be set among
> As, though things be compact of contraries,
> They must by balance have like quantities.

Thus the contrary motions of the heavens imply balance, and 'mild Venus' checks 'fell Mars'. Degree is thus preserved.

But however pessimistic orthodoxy could be about the heaviness of the punishment inflicted through fortune on man for his fall, it always fought the superstition that man was the slave as well as the victim of chance. The classic exponent of this doctrine was for the Middle Ages Boethius, and the Elizabethans accepted him also. The main theme of Boethius's *De Consolatione Philosophiae* is the power of man to survive the blows of fortune. There is no need to repeat the ethical commonplaces found here and in a thousand other places, but the large general contention is that man has it in him to survive the blows of fortune and that ultimately fortune herself is, like nature, the tool of God and the educator of man. The good man ultimately is always happy, the bad man most unhappy when most successful in his evil plans.

Raleigh has a wonderful section on the stars in his *History of the World*, which for the opinion of the educated Elizabethan I cannot do better than summarize. He begins with saying that it is an error to hold with the Chaldaeans, Stoics, and others that the stars bind man with an ineluctable necessity. It is the opposite error to suppose that they are mere ornament.

And if we cannot deny but that God hath given virtues to springs and fountains, to cold earth, to plants and stones, minerals, and to the excremental parts of the basest living creatures, why should we rob the beautiful stars of their working powers? For, seeing they are many in number and of eminent beauty and magnitude, we may not think that in the treasury of his wisdom who is infinite there can be wanting, even for every star, a peculiar virtue and operation; as every herb plant fruit and flower adorning the face of the earth hath the like. For as these were not created to beautify the earth alone and to cover and shadow her dusty face but otherwise for the use of man and beast to feed them and cure them; so were not those uncountable glorious bodies set in the firmaments to no other end than to adorn it but for instruments and organs of his divine providence, so far as it hath pleased his just will to determine.

Allowing not too much and not too little weight to the stars, we may hold that they are not autonomous causes but 'open books, wherein are contained and set down all things whatsoever to come'. But these books are beyond the wit of man thoroughly to read. Even of the plants beneath our feet we know little: how can we then expect to understand the stars above us? But it is difficult to get the question of stars and destiny right, and we must follow a middle course,

that as with the heathen we do not bind God to his creatures, in this supposed necessity of destiny, so on the contrary we do not rob those beautiful creatures of their powers and offices. For had any of these second causes despoiled God of his prerogative, or had God himself constrained the mind and will of man to impious acts by any celestial inforcements, then sure the impious excuse of some were justifiable.

It is undoubted that the stars sway the mind to certain states by acting on our physical predispositions. If a man is weak in will and naturally choleric, for instance, the stars may greatly influence him. Such a man may forget that reason should rule the passions and, prompted by stellar influence, may give way to them. In this he becomes near the beasts, 'over all which, celestial bodies, as instruments and executioners of God's providence, have absolute dominion'. But over the immortal part of man the stars have no necessary sway. 'Fate will be overcome, if thou resist it; if thou neglect, it conquereth.' And there are things to counter the stars' influence, both in nature and in arts. As for nature

> Aristotle himself confesseth that the heavens do not always work their effects in inferior bodies, no more than the signs of rain and wind do not always come to pass. And it is divers times seen that paternal virtue and vice hath his counterworking to these inclinations.

As for art Raleigh is eloquent on the power of education in reinforcing or mitigating the effects of the stars:

> But there is nothing, after God's reserved power, that so much setteth this art of influence out of square and rule as education doth: for there are none in the world so wickedly inclined but that a religious instruction and bringing up may fashion anew and reform them; nor any so well disposed whom, the reins being let loose, the continual fellowship and familiarity and the examples of dissolute men may not corrupt and deform.

The extremes of virtue and vice occur when education or evil communications confirm those whom the stars make naturally virtuous or vicious. Over all is God, who is no more bound to the consistent working of the stars than a monarch to the letter of the law:

> These laws do not deprive the kings of their natural or religious compassion or bind them without prerogative to such a severe execution as that there should be nothing left of liberty to judgement, power, or conscience; the law in his own nature being nothing but a deaf tyrant.

In Elizabethan literature there is a great wealth of references to all possible ways of thinking about the stars, from our being merely the stars' tennis-balls to our faults being not in the stars but in ourselves. But the prevalence of the doctrine that our wills are our own and that the stars' influence can be resisted may not be sufficiently recognized, the typical Elizabethan habit of mind being too often taken to be one of desperate recognition of an ineluctable fate. The series of tragic stories that composes the *Mirror for Magistrates* might easily prompt such a notion, where each character comes with such regularity and hence with such apparent inevitability to ruin. Sackville, too, in his induction constantly refers to these characters as victims of fortune, and it is the stars that help to give the setting of the whole piece:

> Then looking upward to the heaven's leams
> With nightes stars thick powder'd everywhere
> Which erst so glisten'd with the golden streams
> That cheerful Phoebus spread down from his sphere,
> Beholding dark oppressing day so near;
> The sudden sight reduced to my mind
> The sundry changes that in earth we find.

But the true doctrine is uttered by Jack Cade at the beginning of his story:

> Shall I call it fortune or my froward folly
> That lifted me and laid me down below?
> Or was it courage that me made so jolly,
> Which of the stars' and body's 'greement grow?
> Whichever it were, this one point sure I know,
> Which shall be meet for every man to mark:
> *Our* lusts and wills our evils chiefly work.
>
> It may be well that planets do incline
> And our complexions move our minds to ill,
> But such is Reason that they bring to fine
> No work unaided of our lust and will;
> For heaven and earth are subject both to skill.
> The skill of God ruleth all, it is so strong;
> Man may by skill guide things that to him long.
>
> Though lust be sturdy and will inclined to nought
> (This forc'd by mixture, that by heaven's course)
> Yet through the skill God hath in Reason wrought
> And given man, no lust nor will so coarse
> But may be stay'd or swaged of the source;
> So that it shall in nothing force the mind
> To work our woe or leave the proper kind.

In *King Lear* there is a great complexity of reference to fortune and the stars, yet the trend is that of the *Mirror for Magistrates* and of Raleigh; in no way unconventional. Gloucester is foolishly superstitious. Lear is the apparent victim of fortune, yet his 'skill' somehow persists and he is able to carry conviction when he says to Cordelia

> we'll wear out
> In a wall'd prison packs and sects of great ones
> That ebb and flow by the moon.

Edmund's famous satire on superstition is less simple. Gloucester blames the recent eclipses for the present evils in society and gives a typical picture of 'degree' upset:

Love cools, friendship falls off, brothers divide; in cities mutinies; in countries discord; in palaces treason; and the bond crackt 'twixt son and father.

Edmund, when his father goes out, comments satirically:

This is the excellent foppery of the world that when we are sick in fortune (often the surfeit of our own behaviour) we make guilty of our disasters the sun the moon and the stars; as if we were villains by necessity, fools by heavenly compulsion, knaves thieves and treachers by spherical predominance, drunkards liars and adulterers by an enforced obedience of planetary influence, and all that we are evil in by a divine thrusting on: an admirable evasion of whoremaster man, to lay his goatish disposition to the charge of a star! My father compounded with my mother under the dragon's tail, and my nativity was under Ursa Major; so that it follows I am rough and lecherous. Tut, I should have been that I am, had the maidenliest star in the firmament twinkled on my bastardizing.

Here Edmund shows himself the accomplished villain, the man who sins knowingly, who is the ape but not the servant of God. He is as sound as Raleigh on the impiety of allowing 'second causes' to 'despoil God of his prerogative' or of imagining that God 'had constrained the mind and will of man to impious acts by any celestial inforcements', and he is justly the critic of his father's superstitions. But he is brutish in 'robbing those beautiful creatures of their powers and offices'; and brazen in arrogating to himself a viciousness that would have triumphed over every stellar inducement to virtue. However humorous his bogus nativity, we may be meant to look on Edmund as one of those superlatively vicious men whom the stars and their own wills have joined to produce. If this is so, there is dramatic irony in his denying the influence of the stars in words of wickedness that substantiate it in a sense quite other than he intended.

Prospero is the opposite of Edmund, a man in whom reason is strong and who both defies the stars when they are hostile and, when they are kind, uses them to the general benefit. It is possible too that there is an intended connection between the stars and Caliban's insusceptibility to 'nurture' or education. The stars, said Raleigh, had absolute sway over plants and beasts. Caliban's notable feeling for nature may mean a kinship with them in just this subservience. He is too much under the sky's dominance ever to be other than he is.

III. The Elements

Whether or not every educated Elizabethan had it well in his mind that the ether, according to Aristotle, had its native and eternal motion, which was circular, he took the motions and properties of the four elements very much for granted. When Cleopatra said she was all air and fire, the educated part of the audience at least would understand without the slightest effort of memory. The property of air and fire was to go upward in a straight line, as was that of earth and water to go down.

The change of meaning in the word *element* has prevented a modern from understanding the thoroughness with which medieval and modern science are opposed. An element is thought of as an ultimate constituent part, a final result after analysis has done its work; and the four elements are regarded as the rudimentary gropings after an atomic theory instead of something quite opposed. Now just as God, source of all existence, to the medieval mind was first of all one and after was divided in this way or that; so matter was one, and the elements far from being ultimate and different indivisibles were primarily certain qualities attributable to all matter. They were founded on the notions of hot and cold, dry and moist; and earth as an element was the name for the cold and dry qualities of matter in combination. In other words the elements were thought of through their effects. These effects working on a common substance were thought, in cooperation with stellar influence and the occasional extraordinary intervention of God, to explain the way the sublunary world was conducted. But the elements could not be limited to qualities and effects, in ordinary thought. Nemesius puts the ordinary notion plainly enough:

Every one of these elements hath two coupled qualities, which constitute the species or nature of it. Yet these qualities by themselves cannot be elements; for qualities are void of body, and of things incorporeal things corporeal cannot be made. It follows therefore necessarily that every element is a body and a simple-body, and such a one as hath actually in it, in the highest degree, these qualities: heat cold moisture and dryness.

The elements therefore as well as being effects were at least aspects of the common substance, and as such they had their almost ceremonial places in the great world order.

Heaviest and lowest was the cold and dry element, the earth. Its natural place was the centre of the universe, of which it was the dregs. Outside earth was the region of cold and moist, the water. That solid land should thrust itself above the waters was merely one of the many instances of an extrinsic cause making a thing depart from its own intrinsic nature. Outside water was the region of hot and moist, the air. Air though nobler than water was not to be compared with the ether for purity. Just as angels took their shapes from the ether, so the devils took theirs from the air, their peculiar region. Noblest of all is fire, which next below the sphere of the moon enclosed the globe of air that girded water and earth. It was hot and dry, rarefied, invisible to human sight, and was the fitting transition to the eternal realms of the planets. In this region meteors and other transient fires were generated. These, as transient, could not come from the eternal region of the stars.

But though the elements were arranged in this hierarchy, in their own chain of being, analogous to that of the living creatures, they were in actuality mixed in infinitely varied proportion and they were at perpetual war with each other. For instance, fire and water are opposed, but God in his wisdom kept them from mutual destruction by putting the element of air between them, which, having one quality of both the others, acted as a transition and kept the peace. It is this war of the elements that Tamburlaine makes the precedent of his own ambitious strife:

> Nature, that fram'd us of four elements
> Warring within our breasts for regiment,
> Doth teach us all to have aspiring minds.

The finest results came from a proper balance. For durability a thorough compound was necessary. Such natural things as frost and dew, rainbows and meteors, are evanescent because the compound is imperfect. But in oak or in a diamond the compound is much better, and the actual elements have vanished from recognition into substantial forms. The reason why the beasts have shorter lives than man is that the elements are less well mixed in them. They contain more water and less air, and are thus more open to corruption.

As well as fighting while being susceptible of balance the elements were in a constant flux of transmutation, one into the other. This was one of the great manifestations of mutability beneath the moon, part of the general principle of a constant sum perpetually mutable in the position of its parts: the principle implied in Shakespeare's sixty-fourth sonnet:

> When I have seen the hungry ocean gain
> Advantage on the kingdom of the shore,

And the firm soil win of the watery main,
Increasing store with loss and loss with store.

Of all versions of these notions that pronounced by Pythagoras in his speech in the last book of Ovid's *Metamorphoses* was the most famous in Elizabethan days. It is so simple in statement and at the back of so many pieces of Elizabethan literature that it may be quoted to finish the topic in this place. I use Sandys's translation:

> Nor can these elements stand at a stay,
> But by exchanging alter every day.
> Th' eternal world four bodies comprehends
> Ingend'ring all. The heavy earth discends,
> So water, clogg'd with weight; two, light, aspire,
> Depresst by none, pure air and purer fire.
> And, though they have their several seats, yet all
> Of these are made, to these again they fall.
> Resolved earth to water rarifies;
> To air extenuated waters rise;
> The air, when it itself again refines,
> To elemental fire extracted shines.
> They in like order back again repair:
> The grosser fire condenseth into air;
> Air into water; water, thick'ning, then
> Grows solid and converts to earth again.
> None holds his own: for nature ever joys
> In change and with new forms the old supplies.
> In all the world not any perish quite,
> But only are in various habits dight:
> For, to begin to be what we before
> Were not, is to be born; to die, no more
> Than ceasing to be such. Although the frame
> Be changeable, the substance is the same.

It is this interchange of the elements, as worked by the sun and the moon, that is the theme of one of the bitter speeches in *Timon of Athens*:

> The sun's a thief and with his great attraction
> Robs the vast sea. The moon's an arrant thief,

And her pale fire she snatches from the sun.
The sea's a thief, whose liquid surge resolves
The moon into salt tears. The earth's a thief,
That feeds and breeds by a composture stol'n
From gen'ral excrement.

References to the elements in Elizabethan literature are very
many and their imaginative function is to link the doings of
men with the business of the cosmos, to show events not
merely happening but happening in conjunction with so much
else. The effect is usually cumulative and depends more on a
habit of mind than on a few powerful appeals to the imagination. The idea found in the passage from *Tamburlaine*, quoted
above, of the war of the elements is exceptional; it stands
out from other references to the elements and strikes at once.
And it can take on a different colour from what Marlowe gives
it. It can in fact stand for the chaos that ever since the fall of
man threatens to ruin the universe. Lear's first words in the
storm invoke explicitly all four elements in their uproars; and
though these are presented not in abstraction but as manifested
in the concrete natural happenings, basic elemental conflict is
as much a part of his thought as is the actual violence of the
weather:

Blow, winds, and crack your cheeks, rage, blow.
You cataracts and hurricanoes, spout
Till you have drencht our steeples, drown'd the cocks.
You sulphurous and thought-executing fires,
Vaunt-couriers to oak-cleaving thunderbolts,
Singe my white head. And thou, all-shaking thunder,
Strike flat the thick rotundity o' the world.

The elements were also basic to the study of alchemy, but,
technical references apart as in Jonson's *Alchemist*, they have not
here the same general importance as they have as constituents
of the whole cosmos and, as will be seen in the next section, of
man. But the elemental principle of gold is a piece of lore
genuinely commonplace, so taken for granted in its own day

as to escape notice now. Gold was king of metals, the sum of all metallic virtues; and alchemically it was a mixture of the elements in a perfect proportion. The same perfect proportion in the human body caused health. And the *aurum potabile* of alchemy was the link between the perfect metal and perfect health in the patient. It is this alchemical lore that gives point to the scene between Timon and the two courtezans. They clamour for gold, the symbol of health, and Timon in giving it ironically exhorts them to infect mankind with the appropriate diseases.

BOTH: Well, more gold. What then?
       Believe't that we'll do anything for gold.
TIMON: Consumptions sow
       In hollow bones of man; strike their sharp shins
       And mar men's spurring. Crack the lawyer's voice
       That he may never more false title plead
       Nor sound his quillets shrilly. Hoar the flamen,
       That scolds against the quality of flesh
       And not believes himself. Down with the nose,
       Down with it flat.

Less obvious is the full meaning of the golden lads and girls who in the song over Fidele's grave in *Cymbeline* 'as chimney-sweepers, come to dust'. They may be golden for more reasons than one, but one reason is that they are in perfect health, the elements being in them, as in gold, compounded in perfect proportion.

IV. Man

In the chain of being the position of man was of paramount interest. *Homo est utriusque naturae vinculum.* He was the nodal point, and his double nature, though the source of internal conflict, had the unique function of binding together *all* creation, of bridging the greatest cosmic chasm, that between matter and spirit. During the whole period when the notion of the chain of being was prevalent, from the Pythagorean philosophy to Pope, it was man's key position in creation – a kind of Clapham Junction where all the tracks converge and cross – that so greatly exercised the human imagination. Here is the Pythagorean doctrine as preserved by Photius, the Byzantine lexicographer, in his Life of Pythagoras:

Man is called a little world not because he is composed of the four elements (for so are all the beasts, even the meanest) but because he possesses all the faculties of the universe. For in the universe there are gods, the four elements, the dumb beasts, and the plants. Of all these man possesses the faculties: for he possesses the godlike faculty of reason: and the nature of the elements, which consists in nourishment growth and reproduction. In each of these faculties he is deficient; just as the competitor in the pentathlon, while possessing the faculty to exercise each part of it, is yet inferior to the athlete who specializes in one part only; so man though he possesses all the faculties is deficient in each. For we possess the faculty of reason less eminently than the gods; in the same way the elements are less abundant in us than in the elements themselves; our energies and desires are weaker than the beasts'; our powers of nurture and of growth are less than the plants'. Whence, being an amalgam of many and varied elements, we find our life difficult to order. For every other creature is guided by one principle; but we are pulled in different directions by our different faculties. For instance at one time we are drawn towards the better by the god-like element, at another time towards the worse by the domination of the bestial element, within us.

More than two thousand years later Pope described man in the same terms: his doubtful middle state.

> In doubt to act or rest;
> In doubt to deem himself a God, or Beast;
> In doubt his Mind or Body to prefer;
> Born but to die, and reas'ning but to err;
> Alike in ignorance, his reason such,
> Whether he thinks too little, or too much:
> Chaos of Thought and Passion, all confus'd;
> Still by himself abus'd, or disabus'd;
> Created half to rise, and half to fall;

> Great lord of all things, yet a prey to all;
> Sole judge of Truth, in endless Error hurl'd:
> The glory, jest, and riddle of the world!

The vitality of this great commonplace may have waxed and waned in the years between Pythagoras and Pope but it has never been stronger than in the age of Elizabeth. Not only did Man, as man, live with uncommon intensity at that time, but he was never removed from his cosmic setting. It is the combination of these two facts that gives to Elizabethan humanism its great force. This passage from Romei's *Courtier's Academy*, so like in substance to the Pythagorean doctrine, well expresses the ardour with which the Elizabethans contemplated the position of man:

That most excellent and great God, having with all beauty bedecked the celestial regions with angelic spirits, furnishing the heavenly spheres with souls eternal and having replenished this inferior part with all manner of plants herbs and living creatures, the divine majesty, desirous to have an artificer who might consider the reason of so high a work, admire the greatness and love the beauty thereof, in the end made man, being of all worldly creatures the most miraculous. But this divine workman, having before the creation of man dispensed proportionably of his treasures to all creatures and every kind of living thing, prescribing unto them infallible laws, as to plants nourishment to living creatures sense and to angels understanding, and doubting with what manner of life he should adorn this his new heir, this divine artificer in the end determined to make him, unto whom he could not assign anything in proper, partaker of all that which the others enjoyed but in particular. Whereupon, calling unto him he said: Live, O Adam, in what life pleaseth thee best and take unto thyself those gifts which thou esteemest most dear. From this so liberal a grant had our free will its original, so that it is in our power to live like a plant, living creature, like a man, and lastly like an angel; for if a man addict himself only to feeding and nourishment he becometh a plant, if to things sensual he is a brute beast, if to things reasonable and civil he groweth a celestial creature; but if he exalt the beautiful gift of his mind to things invisible and divine he transformeth himself into an angel and, to conclude, becometh the son of God.

Before enlarging on the use the great Elizabethan writers made of this supreme commonplace I must describe in outline the contemporary conception of man's constitution and of his position in creation. It makes a remarkable instance of the continuity of a tradition that the passage quoted from Photius's little Life of Pythagoras can serve as text for most of my description. The Pythagoreans dwelt on man's unique comprehensiveness: he contained in himself samples of all the degrees of creation, excelling in this not only beasts but the angels, who were entirely spiritual beings. But it was not only a matter of including in himself these samples: man's very anatomy corresponded with the physical ordering of the universe. His frame was compounded of the four elements, and on the same principles as was the sublunary world. There is not space to give all the details of this physical correspondence between microcosm and macrocosm, but some of them will naturally occur in the account of man's physical constitution as thought of by the Elizabethans, which must be my next topic.

Man's physical life begins with food, and food is made of the four elements. Food passes through the stomach to the liver, which is lord of the lowest of the three parts of the body. The liver converts the food it receives into four liquid substances, the humours, which are to the human body what the elements are to the common matter of the earth. Each humour has its own counterpart among the elements. The correspondence is best set out in a table.

ELEMENT	HUMOUR	COMMON QUALITY
Earth	Melancholy	Cold and dry
Water	Phlegm	Cold and moist
Air	Blood	Hot and moist
Fire	Choler	Hot and dry

In normal operation all the humours together are carried by the veins from the liver to the heart, a proper mixture of the humours being as necessary to bodily growth and functioning as that of the elements to the creation of permanent substances. The four humours created in the liver are the life-giving moisture of the body. They generate a more active life-principle, vital heat, which corresponds to the fires in the centre of the earth, themselves agents in the slow formation of the metals. This vital heat is mediated to the body through three kinds of spirit, which are the executive of the microcosm. The *natural* spirits are a vapour formed in the liver and carried with the humours along the veins. As such they have to do with the lowest or vegetative side of man and are under the dominion of the liver. But, acted on in the heart by heat and air from the lungs, they assume a higher quality and become *vital* spirits. Accompanied by a nobler kind of blood, also refined in the heart, they carry life and heat through the arteries. The heart is king of the middle portion of the body. It is the seat of the passions and hence corresponds to the sensitive portion of man's nature. Some of the vital spirits are in due course carried through the arteries into the brain, where they are turned into *animal* spirits. The brain rules the top of man's body, and is the seat of the rational and immortal part. The animal spirits are the executive agents of the brain through the nerves and partake both of the body and of the soul.

I will revert to the higher functions of man later. So far the chief concern has been with the humours; and though the account may be over-simplified, it covers most of the assumptions made by the Elizabethans in speaking of the normal working of the body and of its relation to the other parts of creation. When they used the words *temperament* or *complexion* they had consciously in mind the tempering of one humour (or element, as they often said instead) by another, or the intertwining of the humours that was the cause of character. If a man was of a phlegmatic temperament, they meant that the four humours were mixed in a way that allowed phlegm, the cold and moist humour, to be the most emphatic. In Brutus, according to Antony's testimony at the end of *Julius Caesar*, the elements or humours were mixed just right. He was a perfectly balanced man. The subject of Donne's *Anniversaries* was similar:

> She whose complexion was so even made
> That which of her ingredients should invade
> The other three, no fear no art could guess:
> So far were all remov'd from more or less.

But it usually happened that one humour was, even if a little, prominent, giving a man his distinctive mark. On this rigidly physical theory of character the Elizabethans naturally felt themselves very close to the rest of nature and in particular very susceptible to the action of the stars.

Besides these normal conditions of the humours, there were the abnormal. The normal series of changes by which the humours were thought to reach the brain have been described, but abnormally they may ascend straight from the stomach or other abdominal organ to the brain in a vapour, like vapour ascending from the earth to the air, to be distilled into rain. Catarrh comes from these evil vapours. There was also the terrible possibility of a humour not merely existing to excess, as in a perfectly sane man with some marked idiosyncrasy, but going bad. A humour could both putrefy or be burnt with excessive heat. The most famous kind of corrupt humour was the burnt or adust; and *melancholy adust* was the name usually given to it even if it was one of the other humours that had been impaired. It is this melancholy adust, not the mere predominance of normal melancholy, that is the subject of Burton's treatise.

I now revert to man's higher faculties. Like the body, the brain was divided into a triple hierarchy. The lowest contained the five senses. The middle contained first the common sense, which received and summarized the reports of the five senses, second the fancy, and third the memory. This middle area supplied the materials for the highest to work on. The highest contained the supreme human faculty the reason, by which man is separated from the beasts and allied to God and the angels, with its two parts, the understanding (or wit) and the will. It is on these two highest human faculties, understanding and will, that Elizabethan ethics are based.

Man's understanding, though allied to the angelical, operates differently. The angels understand intuitively, man by the painful use of the discursive reason. Again, the angels have perfected their understanding and are replete with all the knowledge they are able to hold. Man, even though he may in the end rival the angels in knowledge, begins in ignorance. What marks man from angel and beast is his capacity for learning: both his 'erected wit' in perceiving perfection and his aptitude for 'nurture' or education in his raising himself towards it. Hence it was that the learning of a Sidney, a Donne, or a Milton was an ethical and religious matter. To learn was to exercise one of the great human prerogatives. But what is it that man should especially learn? The lowest form of understanding concerns our own immediate acts. Though inanimate things are quite shut out from this, fire for instance not knowing that it burns, the animals have got some understanding of their acts. Man is conspicuous, according to Hooker, by seeking perfection through knowledge of things external to himself. Or, as we might put it, one of man's highest faculties is his gift for disinterested knowledge. It was through that gift that he might learn something of God. But there was another subject of understanding which, all were agreed, was paramount; and that was yourself. Here again was a peculiar human task: irrelevant to the angels because they knew themselves already and to the beasts because it was utterly beyond them. Far from being a sign of modesty, innocence, or intuitive virtue, not to know yourself was to resemble the beasts, if not in coarseness at least in deficiency of education. To know yourself was not egoism but the gateway to all virtue. Erasmus has an eloquent chapter on self-knowledge in his *Manual of the Christian Soldier*. It is the great condition of success in the spiritual warfare. For the chief enemy is within ourselves and if we do not understand him we cannot be victorious.

Therefore seeing that thou hast taken upon thee war against thyself, and the chief hope and comfort of victory is if thou know thyself to the uttermost, I will paint a certain image of thyself, as it were in a table, and set it before thine eyes that thou mayst perfectly know what thou art inwardly and within thy skin.

In Sir John Davies's poem the title *Nosce Teipsum* is inclusive of all moral knowledge, even of the immortality of the soul.

It may not be an accident that of the heroes in Shakespeare's four tragic masterpieces two, Othello and Lear, are defective in understanding and two, Hamlet and Macbeth, in will. In *Lear*, the references to his defects of understanding are particularly clear, though how clear cannot be perceived without the contemporary doctrine. When Goneril and Regan discuss their father after he has divided the kingdom, Goneril speaks of Lear's 'poor judgement' and Regan adds 'yet he hath ever but slenderly known himself'. The meaning here is very rich. We are told that Lear is uneducated; he has not grown up: the play will have the painful theme of the education of a man old and hence so set that only the most violent methods can succeed. And we are meant to think of γνῶθι σεαυτὸν, the perennial adage and the specifically human task. At Regan's words the whole context of man in the universe would have been present to the mind of every educated man in an Elizabethan audience, thereby preparing him for Lear's own frenzied questions of the status of man in nature and his kinship with the beasts and the elements.

We must never forget that the Elizabethans thought of the understanding in close relation to the fall of man. The natural thirst for knowledge and wisdom still survives, but the soul's instruments had been impaired and often shirk the labour by which knowledge is obtained.

The understanding then had to sift the evidence of the senses already organized by the common sense, to examine the exuberant creations of the fancy, to summon up the right material from the memory, and on its own account to lay up the greatest possible store of knowledge and wisdom. It was for the will to make the just decision on the evidence presented to it by the understanding. For man's will, like God's and unlike that of natural agents, is free. Fire has no choice in the matter of consuming stubble; smoke cannot choose to go down rather than up. But, in the words of Hooker, from whom I shall quote again

There is in the will of man naturally that freedom whereby it is apt to take or refuse any particular object whatsoever being presented unto it.

The right use of the will is

to bend our souls to the having or doing of that which they see to be good.

It was against the will rather than against the understanding that morality habitually set the appetite or the passions, the product of the heart. They were not always in opposition; yet, their objects differing basically, at times they must be.

The object of appetite is whatsoever sensible good may be wished for; the object of will is that good which reason doth lead us to seek.

It is not in our power not to be stirred mentally by our appetites but it is in our power to translate them or not to translate them into action.

Appetite is the will's solicitor and will is appetite's controller.

It is only by being thoroughly enlightened by the understanding that the will can be victorious in the eternal battle between passion and reason. Specious arguments on the wrong side, too great haste of decision, and custom all persuade us to a wrong choice. Yet they are no excuse, for

there is not that good which concerneth us but it hath evidence enough for itself, if reason were diligent enough to search it out.

If the fall of man had dimmed his understanding, even more had it infected his will. For though it was possible to make a wrong choice through an error of judgement, it was also

possible for the will to be so corrupt as to go against the
evidence of the understanding. One reason why Ovid's words,

> video meliora proboque
> Deteriora sequor,

became a tag was that they expressed the consummation of
corruption caused by the Fall. It is through the will's corruption
that the many diatribes against man, of which the following
from Sir John Hayward is a fair sample, conduct their argument:

Certainly, of all the creatures under heaven which have received being from God, none degenerate, none forsake their natural dignity and
being, but only man. Only man, abandoning the dignity of his proper
nature, is changed like Proteus into divers forms. And this is occasioned
by the liberty of his will. And as every kind of beast is principally
inclined to one sensuality more than to another, so man transformeth
himself into that beast to whose sensuality he principally declines. This
did the ancient wise men shadow forth by their fables of certain
persons changed into such beasts whose cruelty or sottery or other
brutish nature they did express.

To a modern, nurtured on Wordsworthian theories of wise
passiveness and generally inclined to believe in intuitions, the
uncompromising exaltation of the understanding and the will
can be displeasing. But to an Elizabethan the old Platonic and
consistently orthodox opposition between the bestial and the
rational in man, between instinct and understanding, between
appetite and will was starkly real; and Milton's statement of
the effects of the Fall on the minds of Adam and Eve is not
specifically puritanical but would have been accepted as self-
evident by any Elizabethan:

> They sate them down to weep; nor onely Teares
> Raind at thir Eyes, but high Winds worse within
> Began to rise, high Passions, Anger, Hate,
> Mistrust, Suspicion, Discord, and shook sore
> Thir inward State of Mind, calm Region once
> And full of Peace, now tost and turbulent:
> For Understanding rul'd not, and the Will

Heard not her lore, both in subjection now
To sensual Appetite, who from beneath
Usurping over sovran Reason claimd
Superior sway.

Here we have the cosmic setting; the storms in man's minds
compared with storms in the macrocosm: the sharp opposition
between reason and passion: and the physical psychology
implied in the sensual appetite being *beneath*, or in a part of the
body lower than the head, seat of the *sovran Reason*. Moreover,
in the events leading up to the Fall Milton is scrupulously
traditional in his morality. Eve ate the apple largely because she
was unaware of the issues, for the serpent had dulled her
understanding by his flattery and his lies. Unprotected by a
deep understanding she was the victim of an impulse of the
appetite.

The battle between Reason and Passion, the commonplace
of every age, was peculiarly vehement in the age of Elizabeth.
The theological trend of the whole sixteenth century had been
Pauline, and in Paul it is this war, not contemplation or
beatitude, that holds the first place. Early in the century
Erasmus had equated the Reason of the philosophers with
what Paul calls 'sometime the spirit, sometime the inner man,
other while the law of the mind', and the very title of the book
from which this quotation comes, the *Manual of the Christian
Soldier*, indicates the context in which this alliance of qualities
finds itself. Protestants in England of the time of Elizabeth had
made the spiritual war as outlined by Paul into their own most
living mythology; and it had found worthy expression in the
first two books of the *Fairy Queen*. But it was not only through
theology that the struggle was intensified. The Elizabethans
were interested in the nature of man with a fierceness rarely
paralleled in other ages; and that fierceness delighted in
exposing all the contrarieties in man's composition. In
particular by picturing man's position between beast and angel
with all possible emphasis they gave a new intensity to the old
conflict. If in Spenser the conflict is more abstract and theo-

logical, in Shakespeare it is so concrete, so particularized into the objects of creation (and the beasts especially) that we are apt to forget that the abstraction was there. Prospero's direct moralizing about his feelings towards his enemies,

> Though with their high wrongs I am strook to the quick,
> Yet with my nobler reason 'gainst my fury
> Do I take part,

should serve as a corrective, and should remind us that, however transmuted, the conflicts of mature Shakespearean tragedy are those between the passions and reason. But Shakespeare animates these conflicts by stating with unique intensity the range of man's affinities whether with angel and beast or with the lovely or violent manifestations of inanimate nature; in other words by his living sense of man's key position in the great chain of being. It is scarcely necessary to illustrate. *Hamlet* is largely animated by Shakespeare's consciousness of man's being in action like an angel in apprehension like a god, and yet capable of all baseness. In Hamlet's reference to his mother's hasty marriage,

> O heaven, a beast that wants discourse of reason
> Would have mourn'd longer,

the whole context is there. The apostrophe to heaven is more than mere interjection and is meant to bring in man's celestial affinities. Reason, man's heavenly part, has been degraded and he has sunk lower than the beasts themselves. Gertrude's sin is not against human decency alone but against the whole scale of being. Thersites, loathsome himself, is for ever degrading his betters into the similitudes of the beasts. *Lear* and *Timon of Athens* are the two plays that would be especially impoverished if the theme of man's relation to beast was cut out.

To go on illustrating would soon become the same as talking generally about Elizabethan tragedy and in very familiar terms. But there is a famous poem which is central to the present topic and yet which is not usually interpreted in terms of it. This is Donne's *Ecstasy*, a metaphysical love poem indeed, but having as a first concern the mixed constitution of man. Here the topic is not the borderline between man and beast but between man and angel; and Donne uses the well-known mystical experience of the *ecstasis* as his medium. This is how Goodman describes the state:

Though the present condition of man be earthly, made of the earth, feeds on the earth, and is dissolved to the earth, and therefore the soul doth less discover herself by her proper actions than doth the material body; yet it is not unknown to philosophy that there is an ecstasis of the soul, wherein she is carried in a trance, wholly and only intending the intellectual functions, while the body lies dead like a carcass, without breath sense motion or nourishment, only as a pledge to assure us of the soul's return.

It is to this mystical state that the two lovers (like Marvell at the climax in the *Garden*) have arrived; and the common rumination of their two souls that have achieved a union outside their bodies concerns the constitution of man.

> This ecstasy doth unperplex,
>   We said, and tell us what we love;
> We see by this it was not sex,
>   We see we saw not what did move.

Unperplex may combine two meanings. It may mean 'enlighten', but it certainly refers forward to 'that subtile knot which makes us man' and means to unravel the strands binding body and soul. Thus the above lines mean that the ecstasy is an analysis both of man and of love, and that in the light of it we see that sex was not the motive of our love and that previously we did not know what that motive was. In other words the ecstasy promotes the great human function of knowing yourself. Then Donne goes on to establish an analogy between this uncommon union of two souls and the union of various categories in every ordinary soul. But there is one great difference. Ordinary souls

> contain
> Mixture of things, they know not what;

they are novices in the task of self-knowledge. But the lovers'
souls, which together have become a single over-soul,

> know
> Of what we are compos'd and made,
> For th' atomies of which we grow
> Are souls, whom no change can invade;

their self-knowledge, like the angels', is perfected. On the
other hand the angelic state cannot be for ever maintained; the
ecstasis is temporary. Body is a necessary alloy of the soul. It is
also the usual medium between one soul and another, just as
the influence of an immortal being in heaven on the immortal
part of a being on earth is through the medium of the material
air beneath the moon. In the constitution of man there is a
double process.

> As our blood labours to beget
> Spirits as like souls as it can,
> Because such fingers need to knit
> That subtile knot which makes us man:
>
> So must pure lovers' souls descend
> To affections and to faculties
> Which sense may reach and apprehend;
> Else a great prince in prison lies.

The meaning of this is as follows. The humours through the
different spirits, natural vital and animal, are for ever striving
upwards, and the animal spirits, their highest reach, are the
meeting-point, the knot that ties the material and immaterial
parts of man together. If the humours do *their* part in making
this complicated creature, souls must do theirs too: they must
follow the reverse process and consent to descend. Only so will
the meeting-point be reached from above. That is our human
condition. If it is not fulfilled, 'a great prince in prison lies':
the soul cannot use its full faculties.

The poem is jointly an exercise of self-knowledge and an analysis of man's middle state.

Finally, we must ever remember that man's status in the range of creation was never separated from his fall and redemption. The following passage sums up the whole situation of man as tersely as any I know:

If we well consider man in the first estate that God created him, it is the chief and principal of God's work, to the end that in him he might be glorified as in the most noblest and excellentest of all his creatures. But if we consider him in the estate of the general corruption spread all over the posterity of Adam, we shall see him nuzzled in sin, monstrous fearful deformed, subject to a thousand incommodities, void of beatitude, unable ignorant variable and hypocrite ... But if we will consider afterward as being made all new by the immortal seed of God's word ye shall see him restored not only in all his first honours and goods but much more greater; for there whereas sin is poured out for to let and hinder him, the grace of God is more abundantly poured out for to succour him, making him a new creature.

V. Animals Plants and Metals

Many educated Elizabethans would have known Montaigne's subversive conjecture about man and cat: has not the cat as good a right to think that it is playing with the man, as the other way round? But they would not have allowed it, any more than they allowed Copernicus and Machiavelli, to disturb the great outlines of their world picture. They did not doubt that the world and its contents had been made for man, and they were not troubled with any qualms about divine justice if that world was made to suffer through his own fall from grace. Man himself had incurred such vengeance, while the world existed so undoubtedly for his benefit, that the sufferings of nature were both dwarfed by comparison with his own and in themselves were of no consequence. From this it follows that the Elizabethans looked on the lower end of the chain of being mainly in the light of themselves. Its great variety and ingenuity were indeed testimonies of the creator's wonderful power, but its main function was to provide symbols or to point morals for the benefit of man. The ant was a wonderful creation, but the chief thing was that he was there for the sluggard to go to. The bees were wonderfully organized, but the chief thing was that they should 'teach the art of order to a peopled kingdom'.

Nevertheless the lower parts of nature had their necessary function apart from man in filling full the sum of creation, and though their qualities have mostly been touched on already I had better summarize them here in their proper place.

In the scale of creation the beasts excel in sensible capacity. In Gelli's *Circe* the snake, with whom Ulysses has been talking, leaves him to rub his skin against a tree,

a sensation not to be equalled by any that I recollect in your state; because I find the pleasure pure, and without alloy, whereas with you, the sweet is so mixed with the bitter, that the latter is by far the most predominant, and leaves a more lasting impression.

They are also content, unlike man, with the mere necessities of their condition. Lear is in perfect accord with current opinion when he says:

> O reason not the need: our basest beggars
> Are in the poorest thing superfluous.
> Allow not nature more than nature needs,
> Man's life is cheap as beasts'.

And Gelli has the same idea in this dialogue between Ulysses and the mole.

ULYSSES: But one would be glad to have more than one has a mere necessity for.
MOLE: Why? especially if it be not suitable to one's nature. For my part I have no more ambition to surpass the perfection of my own kind than you have reason to wish for the luminous body of a star, or to envy a bird the advantage of a pair of wings.

The beasts eat what is good for them, have the instinct for the right amount of sleep, and keep to their proper seasons of courtship. Goodman compares the instinctive sensuous perception of the beasts with the instinctive intellectual perception of the angels.

> Methinks it should stand with right reason that as sense hath sensible objects, so things themselves should present themselves to the understanding, that the mind should not busy herself to make her own objects intelligible but should only pass judgement and censure. This is the condition of dumb beasts in regard to their sense; this is the state of the angels in regard of their infusion; and this should have been the state of man, were it not that man is fallen.

These then are the proper perfections of the beasts, but there is more water in proportion than air in their constitutions and so they are more liable to decay and a short life.

Plants excel in the faculty of growth. That is the reason why Marvell in his *Coy Mistress* speaks of his love as *vegetable* and as *growing* vaster than empires.

Stones excel in durability, and the best of them are the hardest and the most brilliant, like the ruby and the diamond.

For the borderline between classes the elephant (or lion) and the oyster have already been mentioned as at the top and bottom respectively of the animal creation. Other transitions can be illustrated from one of Donne's best-known poems. In *A Nocturnal upon St Lucy's Day* Donne seeks by every ingenuity to set forth his own nothingness. He is the quintessence of the primordial nothing out of which God made the world: he shares in no type of earthly existence.

> Were I a man, that I were one
> I needs must know; I should prefer,
>   If I were any beast,
> Some ends, some means; yea plants, yea stones detest
> And love.

Beasts have a rudimentary form of understanding, and more developed in the higher mammals. Plants, like the sunflower, have some vestiges of sense-feeling. Even among minerals there is a hierarchy not only of beauty but of vitality. This is Nemesius's version of the doctrine, as before in Wither's translation:

For even in stones (which are inanimate creatures, not having in them, for the most part, so much as a vegetative life) there is otherwise a certain power, making them differ from each other even in their stony properties. But the loadstone seemeth very far to exceed the nature and virtue of other stones in that it both attracts iron thereunto and also detaineth it.

Of the symbolism of animals plants and metals, I must not speak. It has nothing specifically to do with the chain of being and it is a subject so difficult to limit that to treat of it would put this book quite out of proportion.

6

THE CORRESPONDING PLANES

THE world picture so far dealt with was vertical: that of a chain beginning on high with the noblest and descending to the meanest things of creation. But the second picture of the same world was largely horizontal. It consisted of a number of planes, arranged one below another in order of dignity but connected by an immense net of correspondences. It is on one of these correspondences, macrocosm with commonwealth, that Ulysses' speech on 'degree' in *Troilus and Cressida* is built up. The different planes were the divine and angelic, the universe or macrocosm, the commonwealth or body politic, man or the microcosm, and the lower creation. With a network of correspondences between all these Shakespeare must have been familiar, as must Andrew Marvell by the very casualness of his references when in the *Garden* he calls the mind

> that ocean where each kind
> Does straight its own resemblance find.

First the mind is like an ocean because it is microscopic, it contains all the bounty of the seas in little. Second the ocean was supposed to contain a complete store of parallels to whatever existed on earth, after the manner elaborated by Kingsley in the *Water Babies* or by Lewis Carroll describing the insects in the looking-glass world.

This resolution to find correspondences everywhere was a large part of the great medieval striving after unity; it was pushed to extreme lengths by Paracelsus and his like; and it survived in its main outlines past the age of Elizabeth.

Much of the doctrine cannot but appear remote and ingenuous to the modern mind, which is quite unmoved by the numerical jugglings and the fantastic equivalences that delighted earlier generations. Philo, the Jewish theologian who flourished at the beginning of the Christian era, appears from his writings to be a reasonable and civilized person. Yet he explains that God took six days to create the world because the number three stands for the male and two for the female and that through the creative act of multiplying them you get six. An Elizabethan would see no contradiction between this piece of mystical mathematics and Philo's reasonableness. Indeed Ben Jonson, for all his good sense, plays on the idea of male and female numbers in his *Masque of Hymen*:

> And lastly these five waxen lights
> Imply perfection in the rites;
> For five the special number is
> Whence hallow'd union claims her bliss,
> As being all the sum that grows
> From the united strength of those
> Which male and female numbers we
> Do style and are first two and three.

A similar state of mind is shown by Capgrave, a chronicler of English history under Edward IV. He dedicated his chronicle to Edward and ends his dedication thus:

Furthermore yet find I a great convenience in your title that ye be cleped Edward the Fourth. He that entered by intrusion was Harry the Fourth. He that entered by God's provision is Edward the Fourth. The similitude of the reparation is full like the work of the transgression; as the Church singeth in a Preface, because Adam trespassed eating the fruit of a tree, therefore was Christ nailed on a tree. We true lovers of this land desire this of our Lord God: that all the error which was brought in by Harry the Fourth may be redressed by Edward the Fourth.

Not that the state of mind I am describing is extinct. In 1914, when Joffre and French were commanders-in-chief, many people were truly delighted that each name contained six letters and that the last three letters of the first name and the first three of the second were identical. But in Elizabethan days the coincidence would have been felt to be truly portentous. Indeed the amount of intellectual and emotional satisfaction these correspondences then afforded is difficult both to imagine and to overestimate. What to us is merely silly might for an Elizabethan be a solemn or joyful piece of evidence that he lived in an ordered universe where there was no waste and where every detail was a part of nature's plan.

Some of the correspondences, belonging properly to this place, I have inevitably mentioned in writing on order and the chain of being; for many parts of creation cannot help figuring simultaneously as links in the chain and resemblances to something on another grade of creation. Thus a primate in one class of creation must be an important link in the chain as being closest to the class above it and must also correspond to a primate in another class. Suppose for instance that, impressed by its size and its leonine dignity, you called a St Bernard the highest of the canines, you could think of it simultaneously as striving to become a lion and as corresponding in eminence to the diamond and the sun. And treating of the humours I could not avoid mentioning the way they were thought to correspond with the elements.

It is not very easy to survey the great correspondences in any tidy order, because writers will sometimes confine themselves to a pair of phenomena, sometimes run to a whole sequence. Here for instance is a sequence of five. It comes from the *Learned Prince*, the first of *Three Moral Treatises* published by Thomas Blundeville in 1580, a poem giving commonplace advice to the prince to be ruled by reason. The correspondences are between God, the sun, the prince, reason, and justice.

> For justice is of law the end,
> The law the Prince's work, I say.
> The Prince God's likeness doth portend,
> Who over all must bear the sway.
>
> And like as God in heaven above
> The shining sun and moon doth place

> In goodliest wise as best behove
> To show His shape and lively grace,
>
> Such is that Prince within his land
> Which, fearing God, maintaineth right
> And reason's rule doth understand,
> Wherein consists his port and might.
>
> But Plato saith God dwells above
> And there fast fixt in holy saws
> From truth He never doth remove
> Ne swarves from nature's stedfast laws;
>
> And as in heaven like to a glass
> The sun His shape doth represent,
> In earth the light of justice was
> By Him ordain'd for like intent.

But the major correspondences can be arranged in pairs, and I shall proceed on that plan, beginning from the highest downwards.

7

THE CORRESPONDENCES

I. Celestial Powers and Other Creations

THIS correspondence was not greatly used. God is usually in the background sustaining the whole order of creation. But he was sometimes compared with the sun. The author of *Cursor Mundi* explained the doctrine of the Trinity through this correspondence. The sun consists of matter light and heat. Matter corresponds to the Father, light to the Son, and heat to the Holy Ghost. Elizabethans could find the same correspondence in Romei's *Courtier's Academy*. Romei was a Platonist and writes not of the Trinity but of God as mind:

As the sun by his most bright colour is the first visible, first seen and first seeing; so the First Understanding, which is God, most mighty and excellent, with his most glittering shining colour and light essential, is the first intelligible, first understood and first intelligent. The sun by his resplendent light exceedeth all the celestial bodies in beauty: the First Intellect (excepting always if it be lawful to make comparison between the finite and the infinite) through his divine splendour and most glittering light is in the intelligible world of all intellects and beatifullest and most supereminent. And as the light of our material fire in these inferior parts representeth the sun, so is the light of the sun in the world celestial the true similitude of divine light and brightness.

Davies of Hereford makes God correspond to the human soul and the three persons of the Trinity to understanding will and memory.

The angels also are made to correspond with other parts of creation. In Donne's *Ecstasy* the lower order of angels called Intelligences, who managed the spheres, are compared with the souls of men directing their bodies. Elyot gives a triple correspondence of angels, the elements, and man:

And like as the angels which be most fervent in contemplation be highest exalted in glory (after the opinion of holy doctors), and also the fire which is the most pure of elements is deputed to the highest sphere or place; so in this world they which excel other in this influence of understanding and do employ it to the detaining of other within the bounds of reason and show them how to provide for their necessary living, such ought to be set in a more high place than the residue, where they may see and also be seen.

Dionysius the Areopagite would have the ecclesiastical hierarchy on earth duplicate the angelic hierarchy in heaven.

II. *Macrocosm and Body Politic*

It was a commonplace that order in the state duplicates the order of the macrocosm. In the *Homily of Obedience* in the 1547 book of Homilies the two orders are put side by side:

> In the earth God hath assigned kings princes with other governors under them, all in good and necessary order. The water above is kept and raineth down in due time and season. The sun moon stars rainbow thunder lightning clouds and all birds of the air do keep their order.

Shakespeare's 'degree' speech in *Troilus and Cressida* begins with the same notion:

> The heavens themselves, the planets, and this centre
> Observe degree priority and place.

John Norden in his *Vicissitudo Rerum* after speaking of oppositions in the skies, saved from disorder by balance, observes the same in the elements and humours and finally in the commonwealth:

> A body politic or public state
> Hath like dissents, which yet assenting stands.
> The king the subject and the magistrate,
> Noble and base, rich poor, peace and warlike bands,
> Law religion, idle working hands,
> Old young, weak strong, good men and evil, be
> Dislike in parts yet in consort agree.

And in *A Christian Familiar Comfort* he compares the state to the heavens and Queen Elizabeth and her Council to the *primum mobile* or controlling sphere, within whose compass any other motion must be contained. But much more frequent than general comparisons is the special one between the sun, the ruler of the heavens, and the king, ruler of the state. The *roi soleil* is indeed one of the most persistent of all Elizabethan commonplaces, and never was Shakespeare using more familiar material than when in the same speech he said

> And therefore is the glorious planet Sol
> In noble eminence enthron'd and spher'd
> Amidst the other, whose med'cinable eye
> Corrects the ill aspects of planets evil
> And posts like the commandment of a king,

or when in a sonnet he makes the rising sun flatter the mountain-tops with sovran eye. It is of course unthinkable that the new despotism of Tudors and Stuarts should not have exploited and enriched the old correspondence. A magnificent example of such exploitation occurs at the end of Jonson's *Irish Masque*. Here the presence of the King is made to transform the apparent wild Irishmen in their mantles into civilized courtiers as the sun dissolves the chains of winter and fosters the spring. As the Irishmen stand before the King James, a Bard sings to them:

> Bow both your heads at once and hearts;
> Obedience doth not well in parts.
> It is but standing in his eye,
> You'll feel yourselves changed by and by.
> Few live that know how quick a spring
> Works in the presence of a king.
> 'Tis done by this: your slough let fall,
> And come forth new-born creatures all.

During this song the masquers let fall their mantles and discover their masquing apparel. Then they dance forth. After the dance the Bard sings this song:

> So breaks the sun earth's rugged chains,
> Wherein rude winter bound her veins;

> So grows both stream and source of price,
> That lately fetter'd were with ice.
> So naked trees get crisped heads
> And colour'd coats the roughest meads;
> And all get vigour youth and spright
> That are but look'd on by his light.

In the previous reign, the moon as well as the sun was set to duplicate the queen in the heavens, and the moon in the galaxy stood for the queen, Diana or Cynthia, among the luminaries of her Court.

Equally common is the correspondence between disorder in the heavens and civil discord in the state. Ulysses' words are again centrally commonplace.

> But when the planets
> In evil mixture to disorder wander,
> What plagues and what portents, what mutiny,
> What raging of the sea, shaking of earth,
> Commotion in the winds, frights changes horrors,
> Divert and crack, rend and deracinate
> The unity and married calm of states
> Quite from their fixture.

As this topic occurs in some of the best-known passages in Shakespeare, for instance the descriptions of heavenly confusion sympathizing with Caesar's death, no more illustration is necessary. It is characteristic of a later age that Marvell should have turned solemnity into exquisite prettiness. His mower addressing the glow-worms calls them

> Ye country comets that portend
> No war nor prince's funeral,
> Shining unto no higher end
> Than to presage the grass's fall.

Yet for all the prettiness, the delight in finding still one more correspondence persists: and the refinement of seeing the bond between macrocosm and body politic duplicated by glowworm and grass, though not solemn, has its own proper seriousness.

III. *Macrocosm and Microcosm*

Just as in the chain of being the position of man was the most interesting of all, so among the correspondences that between man and the cosmos was the most famous and the most exciting. Most of the details necessary for so short a book as this were given in the course of describing man's physical and mental constitution. It remains to mention some more general or poetical comparisons and to assert that the idea of man summing up the universe in himself had a strong hold on the imagination of the Elizabethans. Sebonde had said that as the noblest heavenly bodies are those highest in the sky, so man's noblest part, his head, is uppermost; and that as the sun is in the midst of the planets, giving them light and vigour, so is the heart in the midst of man's members: and any average Elizabethan would have accepted the sentiment. It is indeed not at all out of harmony with this passage from Raleigh's *History of the World*:

His blood, which disperseth itself by the branches of veins through all the body, may be resembled to those waters which are carried by brooks and rivers over all the earth, his breath to the air, his natural heat to the inclosed warmth which the earth hath in itself . . . the hairs of man's body, which adorns or over-shadows it, to the grass which covereth the upper face and skin of the earth . . . Our determinations to the light wandering and unstable clouds, carried everywhere with uncertain winds, our eyes to the light of the sun and the moon, and the beauty of our youth to the flowers of the spring which in a very short time or with the sun's heat dry up and wither away, or the fierce puffs of wind blow them from the stalks.

For Raleigh is poeticizing something medieval, part of which is more soberly stated in the *Mirror of the World*:

All in like wise as the blood of a man goeth and renneth by the veins of the body and goeth out and issueth in some place, all in like wise renneth the water by the veins of the earth and sourdeth and springeth out by the fountains and wells.

And again:

The sun is the fundament of all heat and of all time, all in such wise as the heart of a man is the fundament by his valour that is in him of all natural heat.

Further from strict physical analogy but based on it is this poem of Fulke Greville, where true love in man corresponds to the eternal light of the fixed stars and of the sun in particular, and lust and its miseries to deprivation of light caused by the earth's own self-interposition:

> Fie, foolish earth, think you the heaven wants glory,
> Because your shadows do yourself benight?
> All's dark unto the blind; let them be sorry:
> The heavens in themselves are ever bright.
> Fie, fond desire, think you that love wants glory,
> Because your shadows do yourself benight?
> The hopes and fears of lust may make men sorry,
> But love still in herself finds her delight.
> Then, earth, stand fast; the sky that you benight
> Will turn again and so restore your glory:
> Desire, be steady; hope is your delight,
> An orb wherein no creature can be sorry;
> Love being plac'd above these middle regions,
> Where every passion wars itself with legions.

Commonest of all correspondences in poetry is that between the storms and earthquakes of the great world and the stormy passions of man. And though here we are often in the realm of metaphor, it is still metaphor strengthened by literal belief. Lear in the storm provides the greatest of all examples. We may go to Greville again, for a cooler and more academic version of the same correspondence, ending with a plea for an escape into a heavenly refuge:

> The earth, with thunder torn, with fire blasted,
> With waters drown'd, with windy palsy shaken,

Cannot for this with heaven be distasted,
Since thunder rain and winds from earth are taken.
Man torn with love, with inward furies blasted,
Drown'd with despair, with fleshly lustings shaken,
Cannot for this with heaven be distasted;
Love fury lustings out of man are taken.
Then, man, endure thyself, those clouds will vanish;
Life is a top which whipping sorrow driveth;
Wisdom must bear what our flesh cannot banish;
The humble lead, the stubborn bootless striveth.
Or, man, forsake thyself; to heaven turn thee:
Her flames enlighten nature, never burn thee.

Morally the correspondence between macrocosm and micro-cosm, if taken seriously, must be impressive. If the heavens are fulfilling punctually their vast and complicated wheelings, man must feel it shameful to allow the workings of his own little world to degenerate. And this is the sense of Hooker's words on this correspondence:

Now the due observation of this law which reason teacheth us cannot but be effectual unto their great good that observe the same. For we see the whole world and each part thereof so compacted that, as long as each thing performeth only that work which is natural unto it, it thereby preserveth both other things and also itself. Contrariwise, let any principal thing, as the sun the moon any one of the heavens or elements, but once cease or fail or swerve; and who doth not easily conceive that the sequel thereof would be ruin both to itself and whatsoever dependeth on it? And it is possible that man, being not only the noblest creature in the world but even a very world in him-self, his transgressing the law of his nature should draw no manner of harm after it?

IV. Body Politic and Microcosm

The best-known instance of this correspondence is in *Julius Caesar*, when Brutus, greatly perplexed, says

Between the acting of a dreadful thing
And the first motion, all the interim is
Like a phantasma or a hideous dream.

> The genius and the mortal instruments
> Are then in council; and the state of man,
> Like to a little kingdom, suffers then
> The nature of an insurrection.

It is a wonderfully full comparison. The commotion in the mind of man, the debate within the highest faculties of understanding and will and the executive faculties such as speech and motion, are likened to a debate of king and council. Just as Plato, to whom the general correspondence goes back, seeks to arrive at individual justice through justice in the state, so Shakespeare illustrates the private by the public commotion. About the time of *Julius Caesar* Davies of Hereford wrote

> And oft it fares in our mind's common weal
> As in a civil war the case doth stand.

There was no need for Shakespeare and Davies to go to Plato for the idea. It had been a commonplace through the ages. Erasmus, for instance, had compared reason in the mind to the king in the state and the mind's low passions to the rabble. Put this way round, with the emphasis on the human passions, the correspondence links with the various allegories of the body and mind, with Spenser's picture of the House of Alma, Richard Bernard's *Isle of Man*, Phineas Fletcher's *Purple Island*. It is preserved in its pure form in Bunyan's *Holy War*, where man and his faculties are presented not as a house but as a city complete in all its parts, political as well as material.

More often the correspondence works the other way; and the usual intention is to establish the unity and the mutually necessary ranks of the body politic, through the correspondence with the human organism. Shakespeare (if it is Shakespeare) puts the thing the usual way round when in the passage already quoted from the *Two Noble Kinsmen* he makes Arcite call Mars the 'shaker of o'er-rank states' and curer of the 'plurisy of people'. Here the overpopulated state comes first, and its cure through war is illustrated by blood-letting in the human body.

In the Middle Ages the correspondence was a persistent

political commonplace, tending to discipline and social stability. It would be as absurd to level social distinctions as to build a body of a number of the same limbs. Here is the doctrine in Elizabethan guise through I. K.'s translation of Romei's *Courtier's Academy*:

Now a city being nothing else but a body of men united together, sufficient of itself to live, it is necessary that like to a human body it be compounded of unlike members, the which, in goodness and dignity among themselves unequal, all notwithstanding concur to the good establishment of a city. Whereupon as it would be a thing monstrous and incommodious to see a human body wholly compounded of heads arms legs or of other members uniform in themselves, so would it be altogether as disproportionable and a thing of itself insufficient if all men in a city were artificers husbandmen soldiers judges or of one self condition and quality.

More positively the different functions in the state are made to correspond to different functions of the body. One of the most elaborate medieval statements is by an Englishman, John of Salisbury, the friend of Thomas à Becket. The notion was still thoroughly alive in the age of Elizabeth. Davies of Hereford embodies his version in the comparison between the Trinity and the human mind mentioned above:

> For as we hold there's but one God alone
> But yet three persons in the Deity:
> So the soul's parted, though in substance one,
> Into understanding, will, and memory.
> These powers or persons make one Trinity,
> Yet but one substance indivisible;
> Which perfect Trinity in unity,
> Both being spiritual and invisible,
> Do make the soul her God so right resemble.
>
> And like as one true God in persons three
> Doth rightly rule this great world's monarchy,
> So in man's little world these virtues be
> But one soul ruling it continually.

Yet in this lesser world, as well we try,
Be sundry sorts of people: some there are
That be as heads, some rulers are so high,
Some common citizens; and some, less rare,
Those rurals be that still [are] out of square.

The heads are those above-recited three,
The under-rulers thoughts and fancies are,
The citizens the outward senses be,
The rurals be the bodies rare
(Which often make the soul most poor and bare);
For when these riff-raffs in commotion rise,
And all will have their will, or nought will spare,
The soul, poor soul, they then in rage surprise,
And rob her of her wealth and blind her of her eyes.

Another Elizabethan version is in Nicholas Breton's *A Murmurer*. This pamphlet is an attack on murmurers against the government and upholds the existing distribution of power. The state is an organism like the human body, and each part of the body must help the others and be helped by them:

God made all the parts of the body for the soul and with the soul to serve him, and all the subjects in a kingdom to serve their king and with their king to serve him. If the head of the body ache, will not the heart be greatly grieved, and every part feel his part of the pain of it? And shall a king in his will be displeased and the heart of his kingdom (the hearts of his subjects) not have a feeling of it? Can the eye of the body be hurt or grieved, and neither the head heart nor any other member be touched with the pain of it? No more can the council, the eye of the commonwealth, be disturbed, but the king will find it and the commonwealth will feel it. Can the hand, the artificer, be hurt but the commonwealth will find the lack of it, the eye with pity will behold it, and the head with the eye, the king with the council, take care for the help of it? Can the labourer, the foot, be wounded, but the body of the state will feel it, the head be careful, the eye searchful and the hand be painful in the cure of it? And can the commonwealth, the body, be diseased, but the king, his council and every true subject will put to his hand for the help of it?

But the most detailed and lively treatment of this theme I know is in Thomas Starkey's *Dialogue between Cardinal Pole and Thomas Lupset*. Starkey was chaplain to Henry VIII at the time Cromwell was chancellor. The *Dialogue* is a beautiful document of the early English Renaissance, full of sweetness and light. It deals with the practical and contemplative life, the law of nature, political theory, and the state of contemporary England. It covers many of the topics of *Utopia* and does not suffer by comparison. Some quotations from it will suffice for all that need be said on this topic. Starkey begins with this general comparison:

Like as in every man there is a body and also a soul in whose flourishing and prosperous state both together standeth the weal and felicity of man; so likewise there is in every commonalty city and country as it were a politic body and another thing also resembling the soul of man, in whose flourishing both together resteth also the true common weal. This body is nothing else but the multitude of people, the number of citizens in every commonalty city or country. The thing which is resembled to the soul is civil order and politic law, administered by officers and rulers. For like as the body in every man receiveth his life by virtue of the soul and is governed thereby, so doth the multitude of people in every country receive, as it were, civil life by laws well administered by good officers and wise rulers, by whom they be governed and kept in politic order.

Starkey then takes various human qualities or conditions postulated by the Greeks for human happiness and applies them comparatively to the state. These qualities or conditions are (1) health strength and beauty (2) friends and wealth (3) virtue. To health of body corresponds the number of citizens. Strength, Starkey says,

standeth in this point chiefly: so to keep and maintain every part of this body that they promptly and readily may do that thing which is required to the health of the whole. Like as we say then every man's body to be strong when every part can execute quickly and well his office determed by the order of nature; as the heart then is strong when he, as fountain of all natural powers, ministreth them with due order to all other; and they then be strong when they be apt to receive their power of the heart and can use it according to the order of nature, as the eye to see, the ear to hear, the foot to go, and hand to hold and reach. After such manner the strength of this politic body standeth in every part being able to do his office and duty ... He or they which have authority upon the whole state right well may be resembled to the heart. For like as all wit reason and sense, feeling life and all other natural power, springeth out of the heart, so from the princes and rulers of the state cometh all laws order and policy, all justice virtue and honesty to the rest of this politic body. To the head, with the eyes ears and other senses therein, resembled may be right well the under officers by princes appointed, for as much as they should ever observe and diligently wait for the weal of the rest of this body. To the arms are resembled both craftsmen and warriors which defend the rest of the body from injury of enemies outward and work and make things necessary to the same; to the feet the ploughmen and tillers of the ground, because they by their labour sustain and support the rest of the body.

Beauty in the body politic consists in the proper proportion to one another of these different classes. Corresponding to friends and wealth of the individual are a full exchequer and friendship with neighbouring states. Crowning all is virtue.

The third, which is chief and principal of all, is good order and policy by good laws established and set, and by heads and rulers put in effect, by the which the whole body, as by reason, is governed and ruled, to the intent that this multitude of people and whole commonalty, so healthy and so wealthy, having convenient abundance of all things necessary for the maintenance thereof may with due honour reverence and love religiously worship God, as fountain of all goodness, maker and governor of all the world; everyone also doing his duty to other with brotherly love, one loving one another as members and parts of one body.

V. General Significance

Although the Elizabethan correspondences are the same as the medieval, they served not quite the same parts of the mind.

The Middle Ages used them more coolly and intellectually, rather as if they were a mathematical formula. They occur fittingly in Chaucer's *Astrolabe*. Higden, the monk of Chester, takes the organization of the universe and all its marvels with a profound serenity. It is not easy to unravel the feelings of the Elizabethans. With their passionate love of ceremony they found the formality of these correspondences very congenial. On the other hand the world they lived in was becoming ever more difficult to fit tidily into a rigid order: the mathematical detail of correspondence became less and less apt; you could not base your faith on the endless accumulation of minutiae. At the same time the desire for order was there. To the correspondence between macrocosm, body politic, and microcosm the Elizabethans gave a double function. On the one hand they made it express the idea of that order they so longed for and on the other serve as a fixed pattern before which the fierce variety of real life could be transacted and to which it could be referred. But they no longer allowed the details to take the form of minute mathematical equivalences: they made the imagination use these for its own ends; equivalences shaded off into resemblances.

This Elizabethan hovering between equivalence and metaphor may become clearer in an example. Modern astronomers, hating the asteroids for being so many and so obstructive, have named them the vermin of the sky. To us this is no more than a metaphor with an emotional content. To the Middle Ages the observation would have been a highly significant fact, a new piece of evidence for the unity of creation: the asteroids would hold in the celestial scale of being the position of fleas and lice in the earthly. The Elizabethans could take the matter either or both ways.

It was through their retention of the main points and their flexibility in interpreting the details that the Elizabethans were able to use these great correspondences in their attempt to tame a bursting and pullulating world. Even if they could not tame a new fact by fitting it into a rigid scheme, at least they could help by finding that it was like something already familiar. If, for instance, the Red Indians could be referred to the men of the traditional Golden Age, their novelty was tamed and they could be fitted into the old pattern and enrich it.

One reason why Ulysses' speech on 'degree' in *Troilus and Cressida* is so rich is that here Shakespeare uses the great correspondences in all possible ways. In the sun-king correspondence he evokes the great pattern of order against which to set the bewilderment of real life; he also uses a poetical metaphor which is yet more than a metaphor. The sun not only *is* the king of the sky but he *is like* the king and the king is like the sun. The great mathematical equivalence and the temporary metaphorical one are simultaneously created. Just the same happens when he says, 'Take but degree away, untune that string.' The abstract world of the great equivalences of the Platonic music is there united with the present world of Elizabethan lute-playing. And when the lute-playing is resembled to the harmony or discord of the political organization, both musical and political activities are made to appear less strange and become part of a more congruent set of worldly phenomena.

8

THE COSMIC DANCE

EVER since the early Greek philosophers creation had been figured as an act of music; and the notion appealed powerfully to the poetically or the mystically minded. As late as 1687 Dryden gave it its best-known rendering in English poetry, keeping strictly to the old tradition.

> From harmony, from heavenly harmony,
>> This universal frame began:
>> When nature underneath a heap
>>> Of jarring atoms lay
>>> And could not heave her head,
>> The tuneful voice was heard from high:
>>> Arise, ye more than dead.
>> Then cold and hot and moist and dry
>> In order to their stations leap
>>> And music's power obey.
>> From harmony, from heavenly harmony,
>>> This universal frame began;
>> From harmony to harmony
>> Through all the compass of the notes it ran.
>> The diapason closing full in man.

But there was the further notion that the created universe was itself in a state of music, that it was one perpetual dance. It was a commonplace in the Middle Ages and occurs in the works of Isidore of Seville, most popular of all medieval encyclopedists. He wrote

Nothing exists without music; for the universe itself is said to have been framed by a kind of harmony of sounds, and the heaven itself revolves under the tones of that harmony.

The idea of creation as a dance implies 'degree', but degree in motion. The static battalions of the earthly, celestial, and divine hierarchies are sped on a varied but controlled peregrination to the accompaniment of music. The path of each is different, yet all the paths together make up a perfect whole. Shakespeare's

> Take but degree away, untune that string,
> And hark, what discord follows,

together with Lorenzo's speech on music shows his knowledge of the general doctrine. Elyot shows the same when he says that the governor's tutor

shall commend the perfect understanding of music, declaring how necessary it is for the better attaining the knowledge of a public weal: which is made of an order of estates and degrees, and, by reason thereof, containeth in it a perfect harmony: which the governor shall afterward more perfectly understand when he shall happen to read the books of Plato and Aristotle of publish weals, wherein be written divers examples of music and geometry.

Like the static notion of degree, the dance to music is repeated on the different levels of existence. The angels or saints in their bands dance to the music of heaven. Milton has his own beautiful version of this dance in *Reason of Church Government*:

The angels themselves, in whom no disorder is feared, as the apostle that saw them in his rapture describes, are distinguished and quaternioned into their celestial princedoms and satrapies, according as God himself has writ his imperial decrees through the great provinces of heaven. Yet it is not to be conceived that these eternal effluences of sanctity and love in the glorified saints should by this means be confined and cloyed with repetition of that which is prescribed, but that our happiness may orb itself into a thousand vagrancies of glory and delight, and with a kind of eccentrical equation be, as it were, an invariable planet of joy and felicity.

Milton speaks poetically not explicitly, but he certainly means that the blessed in heaven resemble the planetary spheres in the variety of their motions and in the music to which those motions are set. Nor is the comparison derogatory to the angels, for of all the dances that of the planets and stars to the music of the spheres in which they were fixed was the most famous.

On the earth natural things, although they shared in the effects of the Fall, are pictured as duplicating the planetary dance. Milton in *Comus*, which as a masque would turn its author's thoughts to dancing, not only expresses his sense of the world's fullness, of the vastness of the chain of being, but passes from immobility to motion. He pictures all the seas dancing in obedience to the moon and allows Comus himself to make the impudent claim that he and his crew are duplicating the dance of the planets:

> We that are of purer fire
> Imitate the Starry Quire
> Who in their nightly watchfull Sphears,
> Lead in swift round the Months and Years.
> The Sounds, and Seas with all their finny drove
> Now to the Moon in wavering Morrice move.

The notion of the seas dancing is not Milton's but is an inherited commonplace. Here is Sir John Davies's* version in *Orchestra*:

> And lo the sea, that fleets about the land
> And like a girdle clips her solid waist,
> Music and measure both doth understand;
> For his great chrystal eye is always cast
> Up to the moon and on her fixed fast.
> And as she danceth in her pallid sphere
> So danceth he about his centre here.

Further than *Orchestra* I need not go, for this poem is the perfect epitome of the universe seen as a dance. Davies published *Orchestra* in 1596 when he was a student at the Inns of Court aged twenty seven. Popularly reputed to be a work of pure

*Not to be confused with John Davies of Hereford, poet and writing-master. Sir John Davies was poet, lawyer, Attorney-General for Ireland, and author of a prose work on the Irish question.

extravagance, it is poetically one of the airiest and nimblest of Elizabethan poems, hovering with accomplished skill between the fantastic and the sublime. It is not for nothing that it is contemporary with *A Midsummer Night's Dream*. In subject matter it combines invention with a mass of cosmic commonplaces. The poet recounts how one night when Penelope at Ithaca appeared among her suitors Athena inspired her with special beauty. Antinous, most courtly of the suitors, begs her to dance or in his own words to

> Imitate heaven, whose beauties excellent
> Are in continual motion day and night.

Penelope refuses to join in something that is mere disorder or misrule, and there follows a *débat* between the two on the subject of dancing, Antinous maintaining that as the universe itself is one great dance comprising many lesser dances we should ourselves join the cosmic harmony. It was creative love that first persuaded the warring atoms to move in order. Time and all its divisions are a dance. The stars have their own dance, the greatest being that of the Great Year, which lasts six thousand years of the sun. The sun courts the earth in a dance. The different elements have their different measures. The various happenings on the earth itself

> Forward and backward rapt and whirled are
> According to the music of the spheres.

The very plants and stones are in some sort included.

> See how those flowers that have sweet beauty too
> (The only jewels that the earth doth wear,
> When the young sun in bravery her doth woo),
> As oft as they the whistling wind do hear,
> Do wave their tender bodies here and there;
> And though their dance no perfect measure is,
> Yet oftentimes their music makes them kiss.

> What makes the vine about the elm to dance
> With turnings windings and embracements round?

What makes the loadstone to the north advance
His subtile point, as if from thence he found
His chief attractive virtue to redound?
Kind nature first doth cause all things to love;
Love makes them dance and in just order move.

In human existence, dancing is the very foundation of civiliza-
tion. Penelope herself is full of the dance without knowing it:

Love in the twinkling of your eyelids danceth,
Love danceth in your pulses and your veins,
Love, when you sew, your needle's point advanceth
And makes it dance a thousand curious strains
Of winding rounds, whereof the form remains,
To show that your fair hands can dance the hey,
Which your fine feet would learn as well as they.

Then, prompted by the God of Love, Antinous in final
persuasion gives Penelope a magic glass to look in, where she
sees first the moon with a thousand stars moving round her and
then the mortal moon, Elizabeth, surrounded by her Court.

Her brighter dazzling beams of majesty
Were laid aside, for she vouchsaf'd awhile
With gracious cheerful and familiar eye
Upon the revels of her court to smile;
For so time's journey she doth oft beguile.
Like sight no mortal eye might elsewhere see,
So full of state art and variety.

For of her barons brave and ladies fair,
Who, had they been elsewhere, most fair had been,
Many an incomparable lovely pair
With hand in hand were interlinked seen,
Making fair honour to their sovereign queen.
Forward they pac'd and did their pace apply
To a most sweet and solemn melody.

Davies never finished his poem, but presumably the sight of
Queen Elizabeth as the central point of the court's dance-

pattern 'in this our Golden Age' would have persuaded Penelope to lay aside her prejudice.

The introduction of Queen Elizabeth and her court is not mere flattery; it shows the cosmic dance reproduced in the body politic, thus completing the series of dances in macrocosm body politic and microcosm. But it stands too for something central to Elizabethan ways of thinking: the agile transition from abstract to concrete, from ideal to real, from sacred to profane. And the reason is the one given before for similar catholicity: the Elizabethans were conscious simultaneously and to an uncommon degree of 'the erected wit and the infected will of man'. It was thus possible for Davies to pass from the mystical notion of the spherical music to the concrete picture of Elizabeth's courtiers dancing, without incongruity.

Orchestra is a fine poem. It serves as pure didacticism, as perfect illustration of a general doctrine. Yet it draws poetic inspiration from the doctrine it propounds. And it has no qualms or doubts about the order it describes. It testifies to the preponderating faith the Elizabethans somehow maintained in their perilously poised world. Not that Davies does not know the things that imperil it:

> Only the earth doth stand for ever still,
> Her rocks remove not nor her mountains meet;
> (Although some wits enrich with learning's skill
> Say heav'n stands firm and that the earth doth fleet
> And swiftly turneth underneath their feet):
> Yet, though the earth is ever stedfast seen,
> On her broad breast hath dancing ever been.

If Davies knew (as here he shows he does) the Copernican astronomy, he must have known that this science had by then broken the fiction of the eternal and immutable heavens. But he trusts in his age and in the beliefs he has inherited and, like most of his contemporaries, refuses to allow a mere inconsistency to interfere with the things he really has at heart.

9

EPILOGUE

IT was fitting to bring the argument of this book to rest in *Orchestra*, for that poem both sums up in itself the commonplaces I have been seeking and is central to the period roughly denominated Elizabethan. It may prompt one to inquire whether there are any general deductions to be drawn from a survey of the Elizabethan world picture. I am inclined to draw three; and that they are as commonplace as the facts from which they originate is nothing against them.

First, the bloom of creative freshness on *Orchestra* together with its so obvious repose in its own age may remind us that the 'real' Elizabethan age – the quarter century from 1580 to 1605 – was after all the great age. Recent attempts to shift the centre of new creative energy to the Metaphysical poets, though intelligible, will not really do. They are like exalting the age of Euripides over that of Aeschylus, or the Perpendicular style of architecture over the Early English. However free one may be to have a personal preference for *Ion* over the *Seven Against Thebes* or St George's Windsor over Wells Cathedral, one is not free to transfer the centre of creation from the earlier to the later work of art. Further, we can estimate the eminence of Elizabethan writers by the earnestness and the passion and the assurance with which they surveyed the range of the universe. By the same means we may find an unexpected kinship between writers too often dissociated. Using this criterion I find that the most eminent are Spenser Sidney Raleigh Hooker Shakespeare and Jonson, and that all these are united in holding with earnestness passion and assurance to the main outlines of the medieval world picture as modified by the Tudor régime, although they all know that the coherence of this picture had been threatened. And these six great writers make Elizabethan literature a massive affair. Milton against unbelievable odds prolongs their spirit in a later age. Donne for all his greatness lacks their assurance; and for him much more than for them the new philosophy called all in doubt. His followers are neat or elegant or florid rather than simple and strong. And it is precisely the basic simplicity and strength of the greatest Elizabethans that we need to perceive if we are not to reduce the norm of their age to mere pageant-making and minstrelsy.

Secondly, the old truth that the greatest things in literature are the most commonplace is quite borne out. I hope that this book has shown to some people how much more common-place than they thought is the substance of some of the writing that appears (and of course in a sense is) most novel and most characteristic of its author. Raleigh's remarks on the glories of creation and on death, Shakespeare's on the state of man in the world seem to be utterly their own, as if compounded of their very life-blood: divested of their literary form they are the common property of every third-rate mind of the age. The rapturous expression apart, Spenser's philosophy is nearly as trite though rather more genteel. The truth is illustrated that the poet is most individual when most orthodox and of his age, *ipsissimus cum minime ipse*.

Finally it must be confessed that to us the Elizabethan is a very queer age. No one can have understood the Elizabethan age who thinks *Orchestra* a poem exceptional to it. It is exactly the poem one would expect from the time; and contemporary readers, however stimulated, must have felt themselves perfectly at home with it. Nevertheless it is a very queer poem, just as the whole substance of this book is a very queer affair. When we are confronted with the notions that God put the element of air, which was hot and moist, between fire, which was hot and dry, and water, which was cold and moist, to stop them fighting, and that while angels take their visible shapes from the ether devils take theirs from the sublunary air, we cannot assume, try as we may, an Elizabethan seriousness. Yet we shall err grievously if we do not take that seriousness into account or if we imagine that the Elizabethan habit of mind is done with once and for all. If we are sincere with ourselves we must know that we have that habit in our own bosoms somewhere, queer as it may seem. And, if we reflect on that habit, we may see that (in queerness though not in viciousness) it resembles certain trends of thought in central Europe, the ignoring of which by our scientifically minded intellectuals has helped not a little to bring the world into its present conflicts and distresses.

NOTES

p. 11. *Nemesius:* Quotation from the *Nature of Man* translated by George Wither, 1636. All subsequent references to Nemesius are from this book.

p. 12. *Shakespeare's England:* This book (2 vols, Oxford, 1917) does indeed contain an excellent chapter on religion, but the true statement made there that 'religion dominated men's minds in the Elizabethan era to an extent not easily comprehended by the modern world' certainly could not be inferred from the other chapters.

p. 16. *Comparison of Elizabeth to the primum mobile:* Theodore Spencer, *Shakespeare and the Nature of Man* (New York, 1942), p. 18, mentions John Case's *De Sphaera Civitatis*, which is prefaced by a diagram picturing Elizabeth as the primum mobile.

p. 22. *Talbot doing homage to Henry VI:* This comes at beginning of *1 Henry VI*, III, 4.

p. 28. *A late medieval theologian:* This is Raymonde de Sebonde, for whom see p. 35.

p. 31. *E. K. Chambers:* I refer to Sir E. K. Chamber's essay, the 'Disenchantment of the Elizabethans', in *Sir Thomas Wyatt and some Collected Studies* (London, 1933).

p. 32. *Raleigh:* Passage is from History of the World, I, 1.

p. 33. *Arthur Lovejoy:* The book is *The Great Chain of Being* by Arthur O. Lovejoy (Cambridge, Mass., 1936). My debt to it is obvious.

p. 34. *Sir John Fortescue:* Works edited by Lord Clermont (London, 1869). I quote with slight alterations the translation by Chichester Fortescue in the above edition, I, p. 322.

p. 36. *Higden:* This author's chronicle *Polychronicon* is best known in Trevisa's translations, but this is very inaccurate and I give my own version. Standard edition in the Rolls series.

p. 36. *Gelli:* For his *Circe* I use the polished eighteenth-century version by H. Layng (London, 1744). Swift was indebted to *Circe* in the fourth book of *Gulliver's Travels.*

p. 37. *Peacham:* The *Complete Gentleman* was published in 1634. It has been reprinted in the Tudor and Stuart Library (Oxford, 1906).

p. 39. *Davies of Hereford:* Passage from *Mirum in Modum* (1602). *The poetical prose-writer* is Sir Thomas Browne.

p. 40. *Lovejoy:* op. cit. p. 165.

p. 42. *W. C. Curry:* In his *Shakespeare's Philosophical Patterns* (Louisiana State University Press, 1937).

p. 44. *Hakewill:* His book first appeared in 1627.

pp. 45ff. *Copernicus, etc.:* For the Elizabethan vogue of the Copernican astronomy through popular books in the vernacular see the early pages of Francis R. Johnson, *Astronomical Thought in Renaissance England* (Baltimore, 1937).

p. 46. *An encyclopedist printed by Caxton:* This is the *Mirror of the World*, translated from the French in 1480 and later printed by Caxton. It is reprinted in the Early English Text Society series. The French originals date from the middle of the thirteenth century. The work was very popular in its French and in its English forms. Among the medieval encyclopedias this is one of the most apt to this book as it is both comprehensive and elementary, containing things universally taken for granted. Quotation is on p. 49 of E.E.T.S. edition, to which future reference is made.

p. 47. *Ether a fifth element:* This is Aristotle's alternative to the Platonic theory that above the moon the elements are perfectly mixed. Plato, *Timaeus*, 31 B; Aristotle, *De Caelo*, 1, 3.
A Frenchman of the time of Francis I: This was Pierre Boistuau, whose *Théâtre du Monde* was translated by John Alday, 1603. Quotation is from the address to the reader.

p. 48. *Sebonde:* Here and in future I use the abbreviation of his *Natural Theology* described on pp. 39–40 above.

p. 52. *Two stanzas from the* Fairy Queen, II, 8, 1–2.

p. 53. *Caxton's encyclopedist:* Passage from *Mirror of the World*, p. 48.
Count Hannibal Romei: His book, which is one of the class of
courtier's manual, was published in 1546. It is prefaced by an
account of life at the court of Ferrara. His cosmology is given in
the first discourse, *Of Beauty.*

p. 54. *Hakewill:* op. cit. II, 2, 2.

p. 56. *Hooker hints at it:* In *Laws of Ecclesiastical Polity,* v 69. All other
references to Hooker are from the first book of this work, for
which references I have not given details of chapters.
Hakewill: op. cit. II, 2, 1 and II, 2, 3.

p. 58. *Burton's long chapter: Anatomy of Melancholy,* I, ii, 1, 2.

p. 60. *It has been well said:* The remark referred to is from W. D.
Briggs's edition of Marlowe's *Edward II* (London, 1914), p. xcv.

p. 61. *General terror of the stars:* For the tyranny of fortune in the early
Christian epoch see the first chapter of Willard Farnham, *The
Medieval Heritage of Elizabethan Tragedy* (University of California
Press, 1936).

p. 62. *Goodman:* op. cit. p. 17.

p. 63. *John Norden's poem:* This is *Vicissitudo Rerum,* 1600. Reprinted
in Shakespeare Association Facsimiles, 1931. Quotation from
stanza 85.
Raleigh: 1, 2, 11.

p. 68. *The Elements:* For medieval and Elizabethan notions of the
elements the writings of Robert Steele are valuable. See *Shakespeare's England,* I, 462ff., article on alchemy, and *Medieval Lore*
(London, 1907), which consists of extracts from Bartholomaeus
Anglicus edited with notes and summaries by Robert Steele.

p. 71. *Ovid:* Quotation is from *Metamorphoses,* xv, 237–58.

p. 75. *Romei:* op. cit. p. 47.

pp. 76ff. *Man's physical life:* There is an excellent account, to which I
am indebted, of man's physical and mental constitution as conceived of by the Elizabethans in Lily B. Campbell, *Shakespeare's
Tragic Heroes* (Cambridge, 1930), pp. 52ff.

p. 78. *Donne:* From the *Second Anniversary*, 123–6.

p. 79. *Erasmus:* In 1533 Wynkyn de Worde printed an English version of Erasmus's *Enchiridion Militis Christiani*. This version has been reprinted in popular form (London, 1905). Quotation from p. 81 of modern version.

p. 82. *Sir John Hayward:* Quotation from *David's Tears*, p. 168. This work, published in 1623, is devotional, being an expansion in prose of the Penitential Psalms.

pp. 83ff. *Protestant mythology:* For the special mythology the Puritans created for themselves see William Haller, *The Rise of Puritanism* (New York, 1938).

p. 85. *Goodman:* op. cit. p. 42.

p. 87. *The following passage:* From Pierre Boistuau, op. cit., end of first chapter.
Montaigne's subversive conjecture: Raleigh for instance echoes it in the *Sceptic* (Works ed. Oldys and Birch, Oxford, 1829, VIII, p. 551).

p. 89. *Goodman:* op. cit. p. 48.

p. 90. *Nemesius:* p. 11.

p. 95. *Romei:* op. cit. p. 14.

p. 96. *Norden:* I derive my reference to *A Christian Familiar Comfort* from D. C. Collins's introduction to the Shakespeare Association edition of *Vicissitudo Rerum*.

pp. 100–101. *Greville:* The two sonnets are from Caelica, Nos. 16 and 27.

p. 102. *Davies of Hereford:* Quotation from *Mirum in Modum*. Works ed. Grosart, 1878, I, p. 24.

p. 103. *Romei:* op. cit. p. 247.

p. 104. *Breton:* Works ed. Grosart, 1879, II, p. 10.

p. 105. *Starkey:* This dialogue and his letters are printed in the Early English Text Society series, 1878. Quotations from pp. 45ff.

pp. 111ff. *Sir John Davies:* Orchestra is accessible in the *Oxford Book of Sixteenth Century Verse*, and is published in separate form by Chatto & Windus.

INDEX

MORE ABOUT PENGUINS
AND PELICANS

Penguinews, which appears every month, contains details of all the new books issued by Penguins as they are published. From time to time it is supplemented by *Penguins in Print*, which is a complete list of all available books published by Penguins. (There are well over three thousand of these.)

A specimen copy of *Penguinews* will be sent to you free on request, and you can become a subscriber for the price of the postage. For a year's issues (including the complete lists) please send 30p if you live in the United Kingdom, or 60p if you live elsewhere. Just write to Dept EP, Penguin Books Ltd, Harmondsworth, Middlesex, enclosing a cheque or postal order, and your name will be added to the mailing list.

Note: *Penguinews* and *Penguins in Print* are not available in the U.S.A. or Canada

Other Volumes by E.M.W. Tillyard available in the
Penguin Shakespeare Library

SHAKESPEARE'S HISTORY PLAYS

In this major study Dr Tillyard sets Shakespeare's history plays against the general background of Elizabethan thoughts, 'a scheme fundamentally religious, by which events evolve under a law of justice and under the ruling of God's Providence, and of which Elizabeth's England was the acknowledged outcome'.

Part I describes the religious, scientific, and political ideas current in Shakespeare's day and enumerates the historical and literary sources for the plays. In Part II the author examines the individual plays in two main tetralogies, with *King John* and *Macbeth* – 'the epilogue of the Histories' – handled in separate chapters.

SHAKESPEARE'S PROBLEM PLAYS

Which are Shakespeare's problem plays? Dr Tillyard argues that there are four. *Hamlet, Troilus and Cressida, All's Well That Ends Well*, and *Measure for Measure*. Several common factors make these into a distinct group: in each of them Shakespeare is acutely interested in speculative thought and in the observation of human nature for their own sake; each tells of men on the verge of manhood and of the harsh experiences which force them to grow up. After discussing these general characteristics, Dr Tillyard traces their effect in the individual plays and so establishes an illuminating relationship between the four.

Shakespeare's Problem Plays forms a scholarly companion to *Shakespeare's History Plays* and *The Elizabethan World Picture*.

'Dr Tillyard's brilliant analyses of these plays do much to illumine what has hitherto been obscure – *Times Educational Supplement.*

Not for sale in the U.S.A.